Institutional Change in Water Management at Local and Provincial Level in Uzbekistan

European University Studies
Europäische Hochschulschriften
Publications Universitaires Européennes

Series IV
Geography

Reihe IV Série IV
Geographie
Géographie

Vol./Band 26

PETER LANG
Bern · Berlin · Bruxelles · Frankfurt am Main · New York · Wien

Kai Wegerich

Institutional Change in Water Management at Local and Provincial Level in Uzbekistan

PETER LANG
Bern · Berlin · Bruxelles · Frankfurt am Main · New York · Wien

Bibliographic information published by Die Deutsche Bibliothek
Die Deutsche Bibliothek lists this publication in the Deutsche Nationalbibliografie; detailed bibliographic data is available on the Internet at ‹http://dnb.ddb.de›.

British Library and Library of Congress Cataloguing-in-Publication Data:
A catalogue record for this book is available from *The British Library*, Great Britain, and from *The Library of Congress*, USA

ISSN 0721-3321
ISBN 3-03910-467-5
US-ISBN 0-8204-7156-9

Hochfeldstrasse 32, Postfach 746, CH-3000 Bern 9, Switzerland
info@peterlang.com, www.peterlang.com, www.peterlang.net

Abstract

The dis-integration of the Soviet Union has led to changes in agricultural and natural resource management in the Central Asian successor states. These changes affected not only the international and national levels of governance but also the provincial and local level governance of the different republics. The focus of the research is on institutional change at the local level in the agricultural sector in Uzbekistan. The study area is located in the downstream regions of the Amu Darya basin, in the Khorezm province and the Karakalpakstan republic. Since the year 2000 a small number of state and collective farms have been privatised and farm organisations and water users associations have been established.

The study deploys political theory, social theory and new institutional economics to analyse and examine institutions and institutional change. Particular attention is given to key stakeholders in the agricultural sector and in water management organisations at the local and provincial levels and to how they implemented the change.

The main findings of the thesis first, confirm that institutions reduce negotiation and discourse costs by providing a structure for contention and accommodation. However, the findings indicate that institutions also have negative effects, because they reflect and reinforce old power relations and therefore can be exploited to serve the interests of privileged former stakeholders. The second main finding is that institutions and their ability to change is determined by the power of influential stakeholders, who enable or reject change.

The findings of the thesis are important for those developing strategies for the creation of bottom-up movements, and also for the social engineering and reengineering of larger institutions and organisations. The findings will be especially significant for those involved in land and water reform processes and in the privatisation of governmental industries.

Abstract

Acknowledgements

I would like to thank my supervisor Tony Allan for his increasing guidance, support and friendship during the research period. Others who have taken the time to provide feedback and comments at the various stages of this work include Waltina Scheumann and Mehmood ul Hassan.

Institutions that served as hosts during the fieldwork were the International Water Management Institute (IWMI), the CGIAR programme facilitating unit in Tashkent and the Institute on Water Resources of the Academy of Science Nukus Branch. I am very grateful to Vilma Horinkova, S.A. Prathapar, and David Seckler from IWMI and Kuatbay Bektimirov from the Academy of Science. The Central Asia Scientific Research Institute for Irrigation (SANIIRI) and the Scientific Information Centre of the Interstate Commission for Water Coordination (SIC ICWC) helped to facilitate the research. I am also grateful to the University of London Central Research Fund, which contributed financially to the research in Karakalpakstan.

I would like to show appreciation to Shirin Akiner and Deniz Kandiyoti from the School of Oriental and African Studies (SOAS) for their support, as well as other scholars and practitioners for their help and support during the research especially Ton Lennaerts, Michael Glantz, Jeremy Berkoff, Markus Mueller, Stuart Horsman, Alisher Ilkhamov and Vadim Sokolov.

I am very grateful to Ed Harris, Daniel Stevens and Adema Joldoshbekova for their hospitality and kindness during my fieldwork in Uzbekistan and Kyrgyzstan.

Also I would like to thank the SOAS water issue group for their inspiration, discussions and support. Especially Anthi Brouma, Itay Fishhendler, Anders Jagerskog, Gerhard Lichtenthaler, Seungho Lee, Geraldine Dalton, He-young Kim and Yasir Mohieldeen.

The thesis would not have been materialised without the support of my friends and family. I am indebted to Urooj Amjad, Kevin Coutinho, Patrick Bangert, Elena Moreno, Manos Markoulatos and Kahin Mariano. I am deeply grateful to my brother Ingo Wegerich and to my mother Traute Wegerich.

Contents

Figures, tables, maps

Figures

Tables

Maps

Abbreviations

BVO	River Basin Organisation
CGIAR	Consultative Group on International Agricultural Research
ICWC	Interstate Commission for Water Coordination
IWMI	International Water Management Institute
NGO	Non-governmental Organisation
NRMP	Natural Resource Management Project
SANIIRI	Central Asia Scientific Research Institute for Irrigation
SIC	Scientific Information Centre
TACIS	Technical Assistance to the Commonwealth of Independent States
USAID	United States Agency for International Development
USSR	Union of Soviet Socialist Republics
WARMAP	Water Resources Management and Agricultural Production
WB	World Bank
ZEF	Centre for Development Research

Chapter one: Introduction

1 Purpose

The purpose of the thesis is to analyse the dynamics of institutional changes of water management at the local administrative level in Uzbekistan. It is argued that it is not possible to analyse the changes at the local administrative in isolation, but that actors at other administrative levels have to be taken into consideration. These actors at higher administrative levels influence the institutional change at the local administrative level.

2 The structure of the thesis

The thesis is divided into the introductory chapter, four main chapters and a concluding chapter. The first chapter introduces the case study and states the significance of the thesis. In the second chapter the theoretical framework is reviewed, The nature of institutions is clarified and a wide range of institutional, social and risk theory is discussed. The scope of institutional change, the influence of elites on institutional change and the motives of elites in influencing institutional change are outlined. The purpose of this rather comprehensive review of theory relating to changing structures and the impacts of participating actors is to identify key ideas of particular relevance to the study.

In order to understand the nature of change in the study area it was necessary to establish the context in which institutional change –

in this case the introduction of water user associations – took place. Therefore, in the following chapter the theoretical framework the 'irreduceable present' as well as the 'inescapable past' of the region is outlined. This discussion is structured into a geographical and into an historical analysis of agriculture and water management in the region from the mid-nineteenth century to the present.

The case study of the establishment of water user associations in Uzbekistan is split into two different administrative levels, the local level, which is described and evaluated in chapter four and the provincial level, which is described and evaluated in chapter five. The chapter on the local administrative level starts with the first land reforms of 1992, which did not incorporate water management reforms. The second part of the chapter focuses on the second land reform, which included the establishment of farm organisations and water user associations on the territories of the former state and collective farms.

Chapter five describes and analyses the influence on the reforms of the actors participating at the provincial and local administrative levels. The motivation of the actors to influence these reforms is examined. The concluding chapter, focuses on the lessons learned from the case study in understanding institutional change and elite behaviour in periods of institutional reform.

The main findings of the thesis are:

1. Through land reforms parts of the elites or actors who belong to the networks of the elites established farms at the local level. These actors had and still have networks to higher levels than the local administrational levels and therefore have more power then the 'normal' local level actor. The new power actors at the local level change the existing balance of power at the local level. The local level as a whole becomes more influential and its bargaining power increases. This implies that it could take action and contradict the interests of the old local elite.
2. Land reform and the introduction of water user associations were in the case study area top-down changes. Through the

land reforms parts of the elite moved to the bottom. Because of the new power relations at the local level, there has been a shift from top-down to bottom-up changes. The new local power actors have demands and have the bargaining power to ask for changes and to implement changes.

3. Whether the new power actors merge with the other local actors is context specific. In the case of the water user associations in Karakalpakstan and Khorezm province, merging took place if all the local actors shared the same resource (here the same irrigation canal) and it depended on the extent to which water scarcity was experienced. In the case where the water scarcity was very severe collective action did not occur, however, when the scarcity was less severe collective action took place. Hence, the success of implementing changes is context specific.
4. The case study demonstrated that different higher administrative levels can function as facilitators for changes at lower levels. In Khorezm, the head of water distribution on the province level tried to facilitate changes on the local level. Facilitation took place in the form of knowledge management, which increased the transparency of the changes and helped to empower the local actors. The higher level facilitation rebalanced in the power relations at the local level between elites and other actors at the local level. The local elites and actors began to have a voice at higher administrative levels. In Karakalpakstan, the head of water distribution on the province level started to see himself as a facilitator.
5. Not every elite was identical. Elites had evolved through different structures. The evidence from the case studies demonstrated that the power of an elite can be based on the patronage system, the political party system or on the educational system. (Other systems have different elites, for example religious elites and market elites). Because the power of the elites was based on different structures, the different elites expressed varying levels of openness and adaptiveness to

changes. In addition, not all changes were equally significant. Therefore, not every change challenged the existing structures equally. Hence, it was possible to get support for different changes from different elites.

3 Background

The break-up of the Soviet Union had, in terms of water management, two major consequences for the Central Asian successor states. Firstly, it shifted the management paradigm from basin-shed management to national administrational-shed management. The second shift happened on the local level, and incorporated complete or partial privatisation of the state and collective farms and the introduction of water user associations on the territory of the privatised farms.

Before independence the Amu Darya river with its Soviet riparian republics Kyrgyzstan, Tajikistan, Uzbekistan and Turkmenistan and the Syr Darya river with the riparian republics Kyrgyzstan, Tajikistan, Uzbekistan and Kazakhstan were managed according to basin boundaries. One centre was responsible for the overall management of the flow. One centre, which could co-ordinate the water to achieve one goal, set by the centre. In addition water management amongst the riparian republics was interlinked with energy and food production. Independence and the emergence of five sovereign states led to a dis-integrated management of resources, such as water, energy and food products. (Compare Wegerich 2002a) With independence the republics focused on national security, which led to a shift from food sovereignty[1] of the republics to attempts of food self-

---

1 'Food sovereignty is one alternative model. It introduces the element of national decision making into food security. It does not demand self-sufficient production at the national level, but instead puts the emphasis for

sufficiency. (compare Wegerich & Bektimirov 2001) Because of the break-up a unified approach to achieve one goal shifted to five national approaches with five goals, which were or are not necessarily different. Given the problem of common pool resources the achievement of a goal upstream might have a negative impact on the possibility of the downstream countries to achieve their goal.[2]

The local level reforms differed in the five Central Asian Republics. While Kyrgyzstan and Kazakhstan opted to reform the agrarian sector towards a market-orientated economic trajectory, Uzbekistan and Turkmenistan chose a path of tight governmental controls, which kept the state ordered system of the former Soviet Union in place for the major crop production sectors of cotton, rice and wheat. While in Kyrgyzstan the agricultural land was completely privatised in Uzbekistan and Turkmenistan the policy on privatisation has been slow. (For Turkmenistan compare Brooks and Lerman 2000) The Central Asian case of irrigation management transfer and the establishment of water user association coincided with land reform. The simultaneous land reform programme made the Central Asian experience different from other adoptions of irrigation management transfer and the creation of water user associations.

In all the republics, which had some form of reform, the reforms were top-down initiated. These reforms anticipated a later shift from a top-down to a bottom-up managed system. Hence, it would appear that the local setting experienced radical institutional change, which should have been associated with a change in the power structure at the local level. For water management this would imply that the anticipated shift should have changed a top-down command administrative or supply driven water management system (in which the hydrologist makes the decision on who becomes water)

---

determining food policy back to the national level, allowing national policy to decide on self-sufficiency or some trade in food as the final outcome.' (Murphy 1999, p. 8) The food sovereignty approach is similar to the livelihood approach for the household but for the national level.

2 For a debate on upstream downstream conflicts along the Syr Darya please see Wegerich 2002a.

to a participatory local empowered demand driven water management system (in which the community of farmers agree on a rotation system, and trade time and water allocation).

Consultant reports and consultant experience from Kyrgyzstan and Kazakhstan indicate that the implemented land reforms were only partially successful, however. The old power structures prevailed and land reforms were manipulated. (for Kyrgyzstan see Baumann 1999 and SERA report 1999, for Kazakhstan see Burger 1998) Similar observations have been made in Uzbekistan during the first land reforms after independence. (Ilkhamov 1998, Egamberdi et al, Eckert & Elwert 2000, Basis Report 2001) The first land reforms are analysed explicitly in Chapter four. The first land reforms in Uzbekistan did not change the water management on the former state and collective farms. Shifts in water management only started with the second land reform in 2000, during which water user associations were introduced.

In semiarid countries land without access to water has only limited use for agriculture. In these countries land and access to water are interwoven and cannot be separated. It is necessary to analyse land reforms and water management in conjunction. The reports on water user association establishment in Kyrgyzstan give evidence about the entanglement between land allocation, access to water and influence on decisions over water resources. The reports of Johnson III (1999) and SERA (1999) show how former stakeholders, former state and collective farm managers and state and collective farm board members, have influence as newly elected water user association board members in the newly established water user associations.

While the literature on Uzbekistan's first land reform entails references to water distribution on the local level, it does not incorporate water user associations establishments. There is only limited data available on the second reforms and the introduction of water user associations in Uzbekistan. (compare Wegerich 2001) The findings of Wegerich are very relevant, because it is the first study on water user associations in Uzbekistan. However the findings are limited, because the report is only based on two water user associations in Uzbekistan, of which one was a European Technical Assistance to the Commonwealth of Independent States (TACIS) pilot project. It is questionable whether the pilot project provides information about old power structures and how the old structures influences changes. In the case of the pilot project outsiders established the association and were responsible for creating a participatory organisation. However, this is not the case for the other established water user associations in Uzbekistan. There, according to Wegerich, old power structures were responsible for the implementation. Wegerich's study indicates that the performance of the two analysed water user associations are very different. He confirms reports from Kyrgyzstan which argue that old structures prevailed in the new organisations. However, because Wegerich's report focuses on water user association sustainability, it does not analyse how and why the institutions 'change' and how the top-down changes are implemented. The research reported in this study attempts to fill the gap on local institutional change and water user association establishment in Uzbekistan.

An important feature of the research is the attention given to the impact of basin levels relations and changes on the management of resources at the local level. The research describes, analyses and evaluates the changes at the local level but emphasises throughout, that the local level and the basin level are connected through the concept of common pool resource management. Inefficient water management upstream, water salinisation and pollution or shifts in the emphasis of crop production upstream, based on either political or economic strategies, will affect water availability downstream. Hence, the dis-integration and the institutional change at the different levels might create short-term or long-term non-favourable or even unsustainable conditions for water user association establishments in particular regions. (compare Wegerich 2002b)

The disintegration of the Soviet Union has had significant impacts on water allocation to the riparian states of Kazakhstan, Kyrgyzstan, Tajikistan, Turkmenistan and Uzbekistan. During the Soviet era the political power determined water allocation amongst riparian states. 'No water has been allocated to Afghanistan. In the agreements of the Central Asian states, it was assumed that Afghanistan would utilise a certain amount.' (Personal communication D.J.W. Berkoff 27.06.2003) The disintegration might change basin water allocation, and might allocate more water to Afghanistan. Such changes in water relations are evident in the case of China, which claimed more water from the river Ile from Kazakhstan. Similar claims could also come from Afghanistan with the stabilisation of Afghanistan along the Amu Darya. In addition water availability in the Aral Sea basin is especially dependent on the wider political setting, for example with the establishment of United States military bases in Uzbekistan talks on water transfers from outside the basin are re-emerging. (Glantz 2002) Hence, even though the research tries to isolate the local institutional change, the isolation is artificial. Agriculture in Uzbekistan is inextricably dependent on irrigation. Changes in water availability and quality at the national level will have an impact on the local setting. However, one has to question whether the local actor is aware of all the changes at the higher levels, and if so, whether he/she acts upon them and changes his/her long or short-term strategy.

4 Significance of the thesis: resources, institutions & theory

The thesis is important because it covers new ground in two different arenas. Firstly, it focuses on institutions and institutional change with an emphasis on different institutional levels and actors influenced by risk and uncertainty. Secondly, it focuses on provincial and local elites during land reform and irrigation management transfer in Uzbekstian. These changes incorporate the shift from state and collective farm to farm organisation and water user association. Thirdly, the local reforms are put into the context of the dis-integration of common pool resource management of the Central Asian Republics in the Soviet era. The dis-integration created not only risk situations for the water management institutions but also had consequences for agricultural production. There was a political paradigm shift that changed water allocation agendas with serious impacts on water resource availability in the downstream regions. Because of the shifts in agricultural priorities the downstream regions experienced severe water scarcity in the 1999–2001 period. Hence the analysis incorporates the wider problems of dis-integration into the analysis of the local level institutional change.

4.1 Why focus on elites and institutional change?

The literature on collective action and on common pool resource management indicates that elites play a crucial role. (Robbins 2000, Baland & Platteau 1999, Shanmugaratnam et al., 1992) What defines these leaders varies in the two case studies. Whether the influence is negative or positive is contested. Das Gupta (2001) argues that hierarchies seriously undermine collective action, however Olson (1971) reasons that inequalities are beneficial in stimulating and sustaining collective action. In both cases the role of elites is important. In addition, it was evident that elites are not only important for common pool resource management, but also play a crucial role in institutional change. The extent to which the influence of the actors is positive or negative depends on which case study is drawn on. (Baland & Platteau 1999, Mosse 1997, Wade 1987) Hence, there are two questions on elites guiding this study. Have elites a positive or a negative effect on collective action? What determines whether elites are supportive or undermining of reform processes?

Is there a key to explain the nature and scope of institutional change? How do institutions change? The literature on institutional change is diverse (North 1990, Binswanger & Ruttan 1978, Douglas 1986). It points to time as an determinant (Sahlins 1987). However, is change determined by 'an inescapable past' or 'an irreduceable present'? Furthermore, change can be marginal or very central and real (North 1990, Douglas & Wildavsky 1982). Institutional change can also be either demand or supply induced. (North 1990, Binswanger & Ruttan 1978) The study attempts to answer the question what is the most appropriate way to induce change and how do such means affect the scope of change?

The evaluation of institutional change and the role of the elites is important for the case study, because it might help to find determinants of the scope of change and the interaction of these determinants during an institutional change process. In addition, it might be possible to deduce some of the results for different cases of institutional change, either in the area of water management or even on a broader level addressing current development challenges. (see main findings at the end of the introduction chapter)

4.2 Theory

While the literature on institutions and institutional change focuses only on one institutional level, the theoretical framework of this study addresses a different perspective by focusing on actors at different institutional levels, and how these actors at different levels interact during change processes.

The literature review on institutions and institutional change (see chapter two) shows that it is possible to distinguish between institutions, which relate either hierarchically or horizontally to each other, in addition one can distinguish between different hierarchical levels of the same institution. The interaction between these institutions and institutional levels is only discussed at the margins, and the existing models are not conducive to, or can explain, top down or bottom up changes and the difficulties or failure of institutional change. The literature review on institutional change shows that changes are influenced through actors, hence it is recognised that the actor is not passive and neutral to the changes but is actively involved.

In addition, the current framework of institutional change utilises the concept of selfishness as primary motivation, 'every individual, though rational, is also selfish' (Downs 1957, p. 27). However not all actions can be explained with this concept. Hence in this research two concepts are adopted, the first is from communication theory (Sperber and Wilson 1986), it is the concept of 'relevance'. The concept incorporates selfish and selfless motivated action. It is argued that institutional change takes place, because of the relevance of the innovation to the actor. The second concept is based on risk, uncertainty and security (see Douglas and Wildavsky 1982). This concept, which could also be incorporated in the concept of relevance, is important because it can be used in a dynamic environment. It is argued that the changes of the system cause new situations for the actor, which he/she interprets as uncertain or secure. The concepts give potential to new interpretations for actions. The concepts incorporate both the actor and the structure approach, because situations are interpreted and reacted upon through structures and through structured solutions.

Another relevant debate, on common pool resources and collective action, indicates that there are different arguments about the influence of elites. (Robbins 2000, Baland & Platteau 1999, Shanmugaratnam et al., 1992) Depending on the argument it is reasoned that elites have either a negative or positive influence. However, the elites are mainly understood as direct stakeholders. However, here elites are not only direct stakeholders at one level, but are identified as actors at different levels. These actors are not directly affected by the changes, but could interpret the changes at a different level as threat, neutral or as opportunity. Because of their interpretations, these stakeholders could get involved and manipulate the situation so that they are secondary beneficials.

4.3 Why Uzbekistan and not Kyrgyzstan?

As mentioned in the beginning of the chapter, the changes in terms of land and water reforms were more rapid in Kyrgyzstan than in Uzbekistan. As early as 1995 in Kyrgyzstan water users formed the first water user associations within the borders of the former state and collective farms. These water user associations were established by individuals. In 1996 the Asian Development Bank (ADB) following a special request by the government carried out work on on-farm irrigation improvement. Workshops were convened, which promoted the importance of forming water user associations. Such workshops have been held in four provinces, namely: Chui, Osh, Issyk-Kul and Tales. This action resulted in an increase in water user associations. In the time period 1996–98 about 50 water user associations were established with a command area of 91,800 ha. By November 1999 about 132,000 ha were covered by 77 water user associations all of them formally established and registered. Besides these additional 130 water user associations were functioning informally on a command area of 125,000 ha. The average size of a water user association is 1,800 ha.

The reports from Johnson III (1999), SERA (1999) and Baumann (1999) indicate that the process of establishing water user associations was not very successful. This situation was further confirmed during the workshop and interviews in Bishkek. According to the results of the workshop, the World Bank (WB) established a scale to measure the success of water user associaitons. Success is measured in seven ‘milestones for irrigation system rehabilitation’. In 2001 in Kyrgyzstan 250 water user associations were established, only 40 had reached milestone four and only 20 milestone seven. Even though, Kyrgyzstan could draw on external help and loans the establishment and sustainability of water user associations had been proven to be difficult. In addition, from the social side, the reports indicated that old local elites had gained during the reforms large proportions of the former state and collective farm resources.

The first field trip to Uzbekistan indicated that the water user association experience in Uzbekistan was just starting and that the process for establishing water user associations was overall slower and not as rapid as in Kyrgyzstan. In addition, the guidelines of establishing water user associations was geared from the Central Asia Scientific Research Institute for Irrigation (SANIIRI), and not from different international organisations. Even though, the European Technical Assistance to the Commonwealth of Independent States (TACIS) had one pilot project, this project seemed to vary tremendously from the other established water user associations. At the first visit only twelve water user associations had been established. The low number and the national level government influence on establishing water user associations in Uzbekistan made water user associations in Uzbekistan as a case study on institutional change at the local level more attractive. In addition, because Uzbekistan is the largest water user in the Aral Sea basin, changes on the local level influencing water use, and the operation and maintenance of the irrigation and drainage systems make the changes in Uzbekistan more significant for the whole basin as changes. In addition, Uzbekistan water user associations have been the most impacted by the interventions of upstream riparian states, which agreed to use only minor portions of the basin waters. (see table in chapter three)

Because of the top-down policy of implementing water user associations in Uzbekistan on unproductive former state and collective farms, the changes in the downstream regions, in the Khorezm province and the Republic of Karakalpakstan, were more rapid then in any other region of Uzbekistan in 2000 and 2001. This tendency towards rapid implementation would continue, and new projects were being planned in other areas, for example in the Kashkadarya province. However, at the time of writing, in Khorezm and Karakalpakstan there were more water user associations in these two regions than in any other regions of Uzbekistan.

The irrigation system in Khorezm and Karakalpakstan has earth canals and open drainage systems, it belongs to the old irrigation systems. (see chapter three). The installation of these old irrigation systems are widespread in downstream provinces throughout the Central Asian river basins, such as Dashkhovuz and Lepab provinces in Turkmenistan and Kyzyl-Orda province in Kazakhstan. The overall objective of the irrigation management transfer projects in the lower Aral Sea Basin – the focus of this study – was and remains the improvement of these old on-farm irrigation systems, which cover an area of more than 3.0 million hectares Rehabilitation is not only very capital intensive, however. Rehabilitation impacts the interests of the newly independent farmers. Hence, an important element of the study will be to show how these institutions change, and whether these changes have been such that the farmers could contribute to the rehabilitation and operation and maintenance of the irrigation and drainage system. One of the key features of water user associations is that in addition to introducing a new participative management structure they provide an institutional context that promotes co-operation and awareness of the benefits of collective action. Water user associations it is argued introduce new incentives and establish or re-establish social capital.

Another contribution of the thesis is its focus on how and why the local and provincial elite influenced the shift from state and collective farm to farm organisation and water user association in Uzbekistan in the years 2000–2001. The study provides a direct account of the reforms which started in the year 2000 and which are still ongoing. Therefore the thesis is useful because it deals with very current material. It also provides a new understanding of the current reforms and how these reforms could be enhanced. Hence, it has a direct policy implication for the ongoing land reforms and irrigation management transfer.

The thesis also provides new material on the current interaction between local and provincial elites in agriculture and water management in a particularly environmentally challenged area. There are historic accounts of local water management in Central Asia, mainly the Ferghana Valley, which describe the changes until the break-up of the Soviet Union (see PhD thesis Thurman, not published). In addition there is the work of Critchlow (1991) on nationalism in Uzbekistan which links national elites to cotton and to external actors, especially Moscow, before the end of the Soviet Union. Recent work (see PhD thesis Weinthal published 2002) links external, in this case international donors, and internal organisations and elite structure after independence to the Aral Sea crisis. But most relevant seems to be Jones Luong PhD thesis (published 2002) on elites on the provincial and national level and the continuity of the political structure after independence. All of these authors, except Thurman with his historic account, tackle elite decision makers at a high administrative level. This thesis tries to evaluate the local level as well as the implementers of decisions and policies of the higher level. The study also shows how these implementers reacted to the top down changes.

The thesis starts where recent articles focusing on the first land reform concluded However, while these articles focus on the first reform and on different elites utilising the reforms to their benefits, this thesis tries to go beyond and to show how the different management and administrative levels interacted after the reform and the advantages and disadvantages of the entanglement between different hierarchies. Ilkhamov's article (1998) on the first land reform shows how the local level manipulates the reforms to their advantage. The study of Eckert & Elwert (2000) already makes the distinction between different actors and looks at the motives for the manipulation of the reforms. However, the articles on the first reform do not show how the story continues after the reform.

5 Methodology of the research

The method of data collection was based on interviews with key informants, individual in-depth interviews, passive observation of group discussions and active participation in group discussions. While the term passive is 'relative', because already the presence of an outsider alters the power positions of the participants within the discussion. Care was taken to try to understand and take into account the biases that could be introduced by discussing the individual encounters with those who had facilitated the field visits. There was a serious attempt to check how the change of power position could have influenced the topics addressed and the willingness of the participants to engage in the discussions. For example during an official meeting in a downstream water user association which suffered severely form water scarcity, the farmers complained that they did not have enough water to fulfil the plan. However, after the official meeting, when the local farmers had left, the chairmen of the farm organisation brought as a gift from his own farm which belonged to the same water user association a water melon. Hence, he did not want to emphasise in front of the other farmers that water scarcity within the same water user association differed. Four main methods of field research were feasible. The extent to which each of these methods was deployed varied according to the perceptions of the interviewee of the interviewer during the 'interviews'. The interviewer was on the one hand perceived as expert/consultant on water user associations or as a research student. Accordingly the interviewer was perceived to be either someone who could be utilised for information, contacts, outside funding and other outside opportunities or someone with an interest in the area but without influence on the current setting. In addition, even though questions were asked about the introduction of water user associations at the local level, the aim of the research was to analyse the behaviour of stakeholders at different levels implementing the changes, and how these stakeholders perceived the changes and whether the changes altered their behaviour or perception of the institutions. Where officials and members from the different administrative levels were present at meetings in the field it was possible to address one of the most important parts of the study, namely how those operating at the different administrative levels interacted. A number of the meetings provided very useful insights into both the nature and direction of the discursive processes relating to water use and management in the downstream irrigated tracts of

Uzbekistan. It was especially interesting to see how the province managers responsible for water distribution in Karakalpakstan and Khorezm argued whether the new independent farmers were sufficiently experienced, and therefore did not need training on irrigation and drainage. It was interesting because it showed that there was not one uniform constructed opinion and it was a new question which challenged the old perceptions of farm employees on the state and collective farms being very specialised.

6 Constraining and enabling factors during fieldwork

6.1 Background material

At the time of writing there was no published literature on the establishment of water user associations in Uzbekistan. In addition, while in Kyrgyzstan water user associations were introduced on a large scale with policy and other support from international agencies such as the World Bank (WB) and the Asian Development Bank (ADB), in Uzbekistan the involvement of the international community on water user association establishment was reduced to the European Union's Technical Assistance to the Commonwealth of Independent States (TACIS) pilot project. Hence, there were neither journal articles nor project reports. In addition, some of the European Union's Technical Assistance officials considered their project to have failed, as a result the available data or information for that project was scarce. (informal interview with Wob Groot, Consultant on water user associations in the second European Union's Technical Assistance to the Commonwealth of Independent States (TACIS) project).

Because the first land reform in Uzbekistan occurred as recently as 1992, the literature is very new and limited. The first literature covering the land reforms focused on economic aspects of the reforms (Lerman 1996, Spoor 1995). Berkoff was the first to argue that the collective farms in Uzbekistan should be transformed into water user associations. (Berkoff 1994) Also because the data were scarce, contradictory statements are common. Literature on the social aspects of the first reform only appeared in the second half of the 1990s (Ilkhamov 1998) or even later. (Eckert & Elwert 2000, Basis Report 2001) It seems that this is because the time-lag between field research and publication. However, because the literature is first generation there has not yet been critical evaluation of this literature nor has there been comparison between the different areas observed and analysed. Unfortunately the published material record similar experience but fail to embed the data in a theoretical framework. The lack of existing research is a problem, it is doubly constraining, because the absence of a theoretical framework makes comparison difficult. The land and water reforms in other Central Asian Republics differ from those in Uzbekistan in timing and content. Further in those republics where reports were available, such as Kyrgyzstan, it could be argued that cotton production was not as dominant as in Uzbekistan, hence the social and cultural setting of the state and collective farms differ.

6.2 Finance

Financial constraints limited the periods spent in the research. Nevertheless, it was possible to visit Uzbekistan twice during the research period, in 2000 and 2001. It was only possible to visit each water user association once. However, a large and sufficient number of water user associations were visited, to enable comparison between the progress made in the different associations and record the role of the elites in the different water user associations. The reaction of the water professional elite to change would not have been possible on the basis of only one water user association case study. However, it was possible to visit Khorezm twice, to observe changes in the approach of the person implementing the establishment of the water user association and training programme.

6.3 Field visits

The first field visit in July 2000 was conducted under the umbrella of the International Water Management Institute (IWMI) which had just started to work on water user associations in Central Asia. Because of its international profile the International Water Management Institute (IWMI) was already connected to the major influential water management institutions based in Tashkent (such as the Central Asia Scientific Research Institute for Irrigation (SANIIRI) and the Scientific Information Centre of the Interstate Commission for Water Coordination (SIC ICWC)). It was possible to derive legitimacy for the field work for this study as a consultant for the International Water Management Institute (IWMI). It was also possible to contact the relevant institutions related to water user association establishments at the national level and the international agency levels and to organise field visits together alongside experts from the Central Asia Scientific Research Institute for Irrigation (SANIIRI) working on the establishment of water user associations. Because of the association with the Central Asia Scientific Research Institute for Irrigation (SANIIRI), detailed information about the technical, environmental as well as institutional background of water user association establishment was readily available. Officials and water user association members at the provincial and local administrative levels proved very willing to participate in interviews. In addition, the accompanying expert from the Central Asia Scientific Research Institute for Irrigation (SANIIRI) could validate and clarify the information collected from the interviews.

During the first fieldwork international and national experts on land reform were interviewed in Tashkent. The interviewees in Tashkent were chosen because of their expertise on water user associations or on land reforms in Uzbekistan. The interviewed experts from the SIC ICWC and SANIIRI had previous contact with IWMI or the Consultative Group on International Agricultural Research (CGIAR), and were therefore recommended to be contacted from IWMI. On the other hand the independent expert from the

Tashkent Centre For Social Research was contacted because of his publications on land reform in Uzbekistan. After consultations with the experts in Tashkent two water user associations were visited. The first one, was the TACIS pilot project in the Syrdarya Province, the second water user association was in the Khiva District of the Khorezm Province. One expert of SANIIRI recommended these two water user associations, because of his familiarity with the two regions and the water user association experience in the two regions.

At the water user association in the Syrdarya Province interviews were conducted with the head of the water user association as well as three members of the board. After the meeting, the farm was visited, and different farmers working on the field were asked about the water user association, and the problems they face with water management.

After an initial meeting with the head of water distribution of the Khorezm Province, it was decided to visit the Khiva district water user association. According, to the head of water distribution of the province level, the Khiva water user association was the most progressive water user association. According to him, it was progressive with training and participation. At the water user association a meeting was initiated with the manager and the board members of the water user association. In the meeting questions were raised on the logistics of the water user association, such as who can become a member of the water user association, number of members, size of farm area of the individual farmers, and what are the different crops produced on the individual land. After establishing the logistical background to the water user association the questioned address other issues, such as what kind of problems are faced by the water user association, what kind of training sessions had been organised and what kind of training sessions would be needed. After the meeting the territory of the water user association with water and irrigation constructions were visited. Farmers, who happened to work on the land during the visit, were asked about the water user association, their awareness of the organisation and whether they received training.

The second field visit during July and August 2001 had institutional back-up from the Institute on Water Resources, Academy of Science, Nukus Branch, Uzbekistan. The link with the Water Resources Institute, and the personnel connection between members of the institute and the Province Manager of Water Distribution of the Province Department for Agriculture and Water Resources in Karakalpakstan as well as one District Manager of Water Distribution immensely facilitated access to these institutions. Furthermore, because of the link with the Academy of Science it was possible to make contacts with other international or national organisations in Karakalpakstan. Because it was the responsibility of the Province Manager of Water Distribution to introduce water user associations into Karakalpakstan he was interested to arrange meetings with water user associations in Karakalpakstan as well as visiting water user associations in the neighbouring province of Khorezm. He was very interested to compare the progress made and to learn from the experience in the other province. Again, because of the association with the Province Manager of Water Distribution it was easier to have access to information on the local level. In addition, because of contact and long discussions it was possible to get background information on his role and his ideas on water user association and how it would facilitate him. Because of his good personal relations with officials in the other province, it was possible to have closer contact with the Province Manager of Water Distribution from Khorezm. It was helpful that the latter had occupied the same position in the year 2000.

Already in the preparation phase in Britain it was planned to visit two water user associations in Karakalpakstan. The emphasis was based on the geographical location of the water user association. Hence, on arrival in Karakalpakistan an upstream and a downstream water user association were chosen to be visited. On both occasions the head of water distribution from the province level as well as the head of water distribution from the different district levels were accompanying the visits. Some members of the water user association were participating in the meetings, however the meetings were dominated by the former state and collective farm manager. During the meeting the first set of questions had the emphasis on the logistics of the water user association, the second part addressed methods and problems of water distribution and whether training sessions had been given and whether they were necessary. After the official meeting, some farms were visited and farmers working on the field were asked about the water user association and problems of water distribution. The visit was closed with an unofficial lunch. During the unofficial lunch more questions were asked on who has access to land and water. On a different occasion a district department for water distribution was visited. The head of the department was a friend of a researcher from the Academy of Science Karakalpakstan branch. Because of the friendship ties, between the researcher who accompanied the visit and the head of the department, it was possible to address issues of corruption and favouritism during the process of land distribution. After the meeting, water management constructions were visited and employees controlling the constructions were questioned.

The typical interview situation during an official meeting with the former manager of the state and collective farm is represented in the diagram below. Only after the official meeting over lunch was it possible to ask questions, which were more critical and which questioned the official versions.

The situation in the diagram shows the authority and the dominating role of the manager of the former state and collective farm. Farmers who were officially invited to the meeting, seemed to have some form of network with the former state and collective farm manager, hence they did not necessarily reflect the average farmer.

Figure 1.1.: Typical interview situation during official meeting

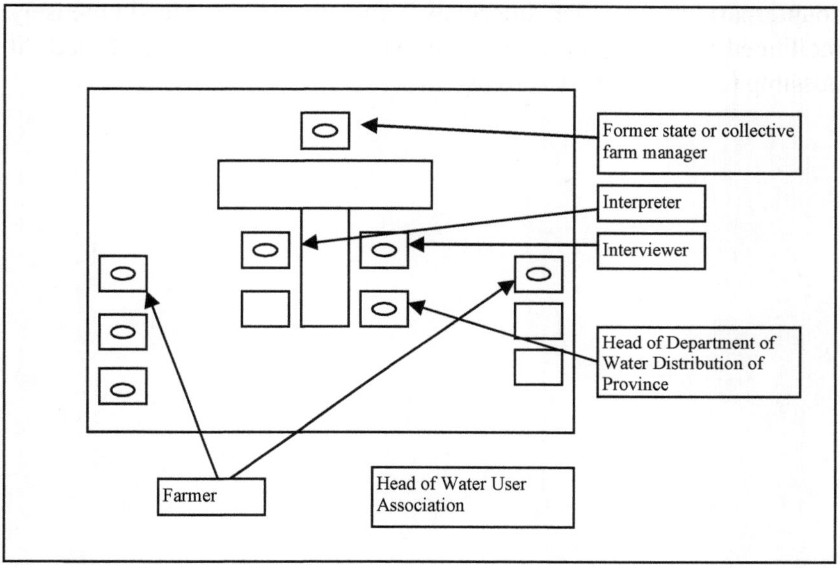

6.4 Language

In most of the interviews it was necessary to rely on an interpreter. Since independence Russian has lost its position as the official language, and Uzbek has become the common language for administration in Uzbekistan and Karakalpak in Karakalpakstan. Russian was only spoken by senior officials. Farmers spoke their local language, either Uzbek or Karakalpak. Because of the need to rely on an interpreter, some meetings became more formal, even though they were unofficial. The interpreters during the first field visit were recommended by the International Water Management Institute's (IWMI) counterpart. During the second field visit, the interpreter was an academic from the Institute on Water Resources of the Academy of Science, who was not a trained interpreter. However, he had worked abroad and was at the same time a local person familiar with local customs. He functioned not only as interpreter but also as local counterpart during the interviews and meetings. He immensely facilitated the field study, provided very useful insights and made it possible to work very effectively.

Chapter two: Institutional change

1 Introduction

This chapter focuses on the links between actor and structure insofar as they influence institutional change. The actor/structure analytical framework enables the integration of different theoretical approaches on institutions, institutional change and the factors influencing change. The theory is important to enable an understanding of the chapters on institutional change at the local and provincial level. On a broader level the results of the analysis are important for those building development strategies for the creation of bottom-up movements, but also for social engineering and re-engineering of larger institutions and organisations. The findings will be especially significant for those involved in land and water reforms and the privatisation of governmental industries.

The analysis starts with an analysis of actor-oriented and structure-oriented approaches. The literature used is based on economics, politics, new institutional economics, anthropology, cultural theory and linguistics. The debate on actor versus structure is influential and therefore common in all of these theoretical frameworks.

It seems that a combination of the actor – structure paradigm enables one the effective analysis of institutions. Institutions will be analysed in terms of their function, their characteristics and their sustainability. These characteristics are significant for the debate on institutional change. A comprehensive understanding of the theory on institutions enables one to understand the prevailing institutions on the local and provincial level in Uzbekistan. The theory can clarify the actions and reactions of the actors during institutional changes. Here, institutional change is analysed from two different approaches: from the supply and the demand side. In the case study of Karakalpakstan and Khorezm the water user associations were supply induced. However, it is important to understand the dynamic between supply and demand induced changes, because even though the water user associations were supply induced, they are supposed to be bottom-up or demand side institutions. The focus on institutions is closely linked to the adaptive capacity of society, which will be explored in the context of the findings on institutional change. The study attempts to demonstrate that both, institutional change and adaptive capacity, are linked to the power[3] of stakeholders. Their role in influencing the institutions and on the process of institutional change will be analysed. In the case of the introduction of water user associations the identified power stakeholders are mainly the former state and collective farm managers on the local level and the province managers for water distribution at the provincial level. Finally, the question whether the notion of uncertainty and risk influence institutional change will be addressed. The focus on risk and uncertainty becomes important because the consequences of disintegration of the Soviet Union and the integrated management of basin water resources triggered different forms of uncertainty for the two analysed administrational levels, the local and the provincial level.

---

3 Power is defined as having control over resources, which could be either natural, financial and labour, but also networks, information, upward and outward mobility. This definition of power incorporates the six capitals: Natural, Physical, Human, Social, Financial and Political. An individual, who is in such a power position is defined as a gatekeeper, because of his access to the 'outside', and his position of control over resource distribution.

2 Institutional theory

2.1 Actor-oriented approach in institutional theory, political science and organisational theory

The actor-oriented approach is based on the assumption that the individual is a rational and selfish actor. Downs, in his economic theory of democracy argues, 'every individual, though rational, is also selfish' (Downs 1957, p. 27)[4]. Downs defines a rational individual as: '(1) he can always make a decision when confronted with a range of alternatives; (2) he ranks all the alternatives facing him in order of his preference in such a way that each is either preferred to, indifferent to, or inferior to each other; (3) his preference ranking is transitive; (4) he always chooses from among the possible alternatives that which ranks highest in his preference ordering; (5) he always makes the same decision each time he is confronted with the same alternatives'. (Downs 1957, p. 6) However, Downs's definition does not take into consideration a changing environment, and therefore the same alternatives could be valued differently in a new context. Awareness of these ideas on variation will be important in the context of the establishment of water user associations and different districts and provinces in Uzbekistan in which water availability varied.

Downs points out that individuals are also socially rational as opposed to individually rational, in other words, socially selfishness and individually selfishness. However, he does not give sufficient explanation of the circumstances in which social or individual selfishness would dominate. Such domination of either form would be important for institution building and institutional change. Because it

4 Berkoff points out 'that "individuals maximise welfare as they conceive it, whether they be selfish, altruistic, loyal, spiteful or masochistic" (Becker's 1993 Nobel Prize speech)' (Personal communication D.J.W. Berkoff 27.06.2003)

would give answers to an actor on when to trust an individual or to act on the basis of social or individual selfishness, and how to motivate an individual to behave in a socially selfish way rather than in an individually selfish manner. In the case of the provincial managers of water distribution in Karakalpakstan and Khorezm without doubt their actions have benefits for the local level, however as will be argued their prime motivation was individual selfishness.

Opp argues that individual and social rational are not necessary contradictory. Opp's approach is based on the assumption that individuals are able to act out of an altruistic motive, which can be interpreted as selfish as well. (Opp 1999, p. 172) However, the distinction between individual and social seems to be inadequate, and a further distinction would be necessary, social has many levels which are horizontally and hierarchically linked. In addition, an action can be defined on one level as individual selfish, but on another level as social selfishness. But these levels do not have to be connected and the actor does not necessarily have to see the links between the two levels. Furthermore, the concept of selfishness as primary motivation does not necessarily explain love and selfless actions. Sperber and Wilson in their analysis of communication come to the conclusion that the understanding of utterances and speech-acts is determined through 'relevance'. (Sperber and Wilson 1986) This implies that the individual chooses a certain option of meaning, because this option is the most relevant to him/her. The choice is time and context dependent. The theory implies that the individual is a rational actor. The theory of relevance enables the explanation of altruistic and selfish motives at the same time. The assumption of relevance – either selfish or altruistic – as motivation for individuals is important on a broader level for institutional theory as well as political science and organisational theory, in the context of this study, relevance becomes important to understand the different approaches of individual actors to reach the same goal. (see Chapter five)

The public choice approach within political theory uses Downs' framework. Mueller defines public choice as 'the economic study of non-market decision making, or simply the application of economics to political science'. (Mueller 1989, p. 1) According to rational choice political theory, the politician in power and the voter are both rational and selfish. The politician within a democratic government acts rationally to maximise his/her support to get re-elected in the following election period. (It is argued that a dictator acts similarly to find support for his/her regime.) This would imply that politicians do not 'seek office as a means of carrying out particular policies; their only goal is to reap the rewards of holding office per se'. (Downs 1957, p. 28) In the context of the rational choice argument of selfishness, altruistic behaviour leading to social functions are only by-products, and the private ambition of the politician is the main goal of his/her actions. This again could be very interesting for the case study on the local as well as the provincial levels, to find an explanation for the actions of the manager of the farm organisations and for the water user associations. In addition Downs' statement of holding office could also be applicable at the higher provincial level, with the idea of the provincial actors get involved in the local level. Similarly, the voter, as an individual, chooses a party according to the highest level of utility income promised. Utility is loosely defined and implies securitisation of livelihoods. This becomes important because of the interpretations why an old stakeholder such as the old elites would have high chances of getting elected or re-elected. The introduction of relevance softens the public choice approach and enables the interpretation of actions as less determined by selfishness.

While in economic theory it was established that individuals employed in firms are utility maximising agents, the theory has not until recently been extended to governmental organisations. The employees of governmental organisations were seen as altruistic motivated individuals, furthering the utility of the society. Downs' theory changed the image of bureaucrats. He used his individualistic selfish-motivated approach to explain the behaviour of bureaucrats, and argued that the individual within an organisation aims to increase his/her income, prestige and power. Hence, the individual within a governmental organisation is not different from the individual in an economic organisation. Niskanen, utilising Downs' approach, argued that a bureaucrat's utility function is influenced by: 'salary, prerequisites of the office, public relation, power, patronage, output of the bureau, ease of making changes and ease of managing the bureau'. (Niskanen 1994, p. 38) The theory of the utility function[5] becomes interesting in the context of the actors at the provincial level and their emphasis on the new institution, which increases exactly these determinants of the function. In the case of Karakalpakstan, the province manager of water distribution wanted to establish an institution, which allowed him to have more power, flexibility and stability.

2.2 *Structure, habits, and institutional oriented approach*

In contrast to the actor-oriented approach is the emphasis on structures, either habitus[6], informal or formal institutions, and their deterministic nature on the behaviour of the actor. Here institutions are defined as frameworks within which human beings interact. 'Institu-

---

5 Niskanen took Downs' approach a step further. He noticed that most of the variables in the utility function were connected to the budget of the organisation. Therefore, he concluded that the bureaucrat's utility is 'positively and continuously associated with the level of the budget'. (Niskanen 1994, p. 38) Niskanen's reasoning, leads to the conclusion that a private firm and a public organisation are comparable: both are interested in growth. While the firm attempts to increase its profits, the bureau attempts to increase its budgets. However, maximisation of the budget is neither related to an optimal output or to efficiency within the organisation. However, the reduction of the utility function through the maximisation of the budget is very restricting, especially if this approach is supposed to be general and not only applicable in the context of modern organisations, but also in pre-modern and reflexive modern contexts.

6 Habitus is the term coined by Bourdieu to capture the influence of established practices in the immediate environment of individuals on their behaviour. In the following, instead of habitus the term habit will be utilised.

tions are a set of rules, compliance procedures, and moral and ethical behavioural norms designed to constrain the behaviour of individuals.' (North, 1981: pp. 201–202, in Feeny 1988, p. 171)

Durkheim argues against the pure individualistic approach, because it leaves unexplained group solidarity, for example of small groups and religious communities. More structuralistic is the approach of Scott and Meyer, who argue that the 'central institutional effects entail not influences on actors but the creation of actors' (Scott and Meyer 1994, p. 5) This implies that the individual is determined by the institutional environment, and can only act within the given structures of his/her environment. Bourdieu (1990) extends the theory of structures and claims that the individual is influenced by the practices of behaviour in its immediate environment. Therefore, the individual is not only influenced by formal structures but also by the practices of actors. Ortner combines practice and structure, she argues that structures 'are grounded in recognisable practices of ordinary life' (Ortner 1989, p. 75).

Structure oriented approaches argue that structures determine the environmental perception of individuals. Hence, Douglas reasons 'institutions systematically direct individual memory and channel our perceptions into forms compatible with the relations they authorise'. (Douglas 1986, p. 92) Structures of the social environment screen the prospects for the individual. (compare: Douglas & Wildavsky 1982, p. 72) Bourdieu reasons, 'as an acquired system of generative schemes, the habitus makes possible the free production of all the thoughts, perceptions and actions inherent in the particular conditions of its production – and only those'. (Bourdieu 1990, p. 55) According to Bourdieu habits produce individuals with attitudes based on the structures generated by history. Sahlins calls it 'an inescapable past because the concepts by which experience is organised and communicated proceed from the received cultural scheme' (Sahlins 1987, p. 151) Consequently, humans are social beings, who have internalised social interaction and even their decision-making processes are determined by structures. The argument of the 'inescapable past' which determines the present becomes important in many different settings of the case study. In the background chapter it was shown how the Soviets tried to implement new structures, which were supposed to erase unwanted old structures. And even in the present the introduction of new structures such as water user associations seem to be threatened by old structures which have 'survived' the Soviet period or are Soviet invented structures. Structuralism has further implications for rational choice approaches. Rational choice is rational only in the context of a structured environment, but can be interpreted as irrational in a different time and space dependent environment. (for example: witchcraft, is rational in a certain institutional environment).

Douglas argues that institutions 'are founded in nature [...] they are part of the order of the universe'. (Douglas 1986, p. 52) Hence, according to Douglas the individual is born into an already existing structure determined and founded by nature. Less deterministic is the reasoning of Bourdieu. He argues that the individual is influenced by the notion of the habits into which he/she is born. The habits are not founded in nature, but constructed through history and past experience of actors. Ortner reasons that the individual is only loosely connected to the structure, and that the individual 'finds most of his or her culture intelligible and meaningful'; however, the individual 'does not necessarily find all parts of it equally meaningful in all times and places' (Ortner 1989, p. 198) This implies that there are different structures existing at the same time and space. These structures can be contradictory. According to Ortner there is a constant challenge between contradictory structures, which the individual has to respond to in standardised schemes, or he/she has to find an alternative, which could lead to another structure. In the context of the land reforms on the local level it was reasoned from the administration that a 'socialist mentality' prevented employees from applying for land. However, according to Ortner, different structures might have been responsible for low turnout. North's approach is more actor orientated, he argues, 'institutions [...] are the humanly devised constraints that shape human interaction'. (North, 1990, p. 3)

While Douglas reasons that institutions are constant, Bourdieu's habitus is able to change over time, however, a habit derives 'disproportionate weight [from] early experience' (Bourdieu 1990, p. 54) On the other hand Ortner's approach is not that clear cut. Because if the individual can value different things at different times and places, then it would be difficult to prove whether a structure is changing or whether the original structure is still in place and used. North's approach is the most flexible. He sees institutions to be changing constantly, indifferent to time and space, and only dependent on human action. Sahlins' approach accords with North's explanation. According to Sahlins the individual is trying to reproduce the structure with every action and with every new event he/she acts upon. However, the personal interest of the individual influences the reproduction of the structure. Sahlins calls this 'an irreduceable present'. 'The difference lies in the irreducibility of specific actors and their empirical contexts, which are never precisely the same as other actors or other situations.' (Sahlins 1987, p. 152) Hence, the sustainability of structures in North and Sahlins approach is constantly challenged. For Douglas institutions would succeed when they are based on natural structures, and according to Bourdieu when they are based on past practices. If structures are new and not related to the existing structures then the transaction costs to keep these new structures in place would be high. The introduction of transaction costs into the debate of the nature of structures and their sustainability leads to an analysis of the functions of institutions. Arguably, the transaction costs of establishing an equitable, participatory, bottom-up organisation in a society, which is predominantly based on top-down and patronage structures, will be much higher than the establishment of such an organisation in a society, which is based on similar values as the organisation. Therefore, it seems likely that the introduction of a participatory organisation such as a water user association faces problems from the members as well as the wider institutional context.

2.3 *The functions of institutions*

The perception of the purpose of institutions varies. Coming from an actor oriented approach, North abandoned the idea that institutions are created to enhance efficiency, and argues, based on a study of historic shift in property rights, that 'rulers devised property rights in their own interests and transaction costs resulted in typically inefficient property rights prevailing'. (North, 1990, p. 7) North's statement might be helpful to understand the reforms on the local level, however North's reasoning is only one dimensional, from top to bottom. The case of land reforms in Uzbekistan will show that there is more than one top, and that these tops are in competition with each other. North points out that, 'institutions reduce uncertainty by providing a structure to every day life'. (North, 1990, p. 3) Hence, institutions can be transaction cost minimising. The approach is taken up by Nabli and Nugent who argue, 'institutions are transaction cost minimising arrangements, which may change and evolve with changes in the nature and sources of transaction costs and the means of minimising them'. (Nabli & Nugent 1989, p. 1336)

Scott and Meyer point out that it is necessary to distinguish economic, political science and sociological approaches. (Compare: Scott & Meyer 1994, p. 2) However, institutionalists like Bromley come to the same conclusion as North. Bromley and Gibbs argue 'Rules, prescribe individual functions within the group and thus eliminate the need of members to negotiate every new transaction with each other'. (Gibbs and Bromley 1989, p. 26) Even anthropologists like Douglas and Wildavsky repeat the transaction cost minimising function. They envision the role of institutions as 'problem-simplifying devices', and reason that individuals 'delegate their decision-making processes to institutions'. (Douglas & Wildavsky 1982, p. 80) The implication is that institutions reduce transaction costs for the individual, because institutions determine the rules of conduct. However, the difference between the approaches of North and Douglas is that the latter sees institutions not just as the creations of individuals to reduce the transaction costs of daily decision-making, but as wider environments which influence and shape the individuals. Douglas' argument is also taken up by Scott and Meyer, who argue that, 'visible structures and routines that make organisations are direct reflections and effects of rules and structures built into wider environments.' (Scott & Meyer 1994, p. 2)

Hobley argues, 'institutional structures are 'capital' passed down from history'. (Hobley 1996, p. 11) (The term capital reminds one of the social capital approach of Putnam (1993), Fine (2001) and the World Bank, see below on adaptive capacity) She continues that they condition power relations within society. 'Institutions determine (a) the power relations between state and citizen and (b) equity distribution between citizens, as well as having an indirect effect on social and economic livelihoods.' (Hobley 2000, p. 11) (This idea is already connected to horizontal and hierarchical structures in society, see below on elites) The notion that an institution is a resource, together with labour, capital and technology is also reflected by Ostrom, Feeny & Picht who argue that law as an institution 'is as much a part of the mode of production of a society as farmland and machinery; farmland or machinery is nothing unless it operates, and the law is an integral part of its operation'. (Ostrom et al. 1988, p. 440) Hence, institutions are social resources, which determine relations of individuals, groups and organisations, furthermore institutions enable or disable 'developments' of society.

As Scott and Meyer argue 'institutionalisation processes can occur at various structural levels – from interpersonal and intra-organisational to organisational field, to societal and even to the trans-societal level' (Scott & Meyer 1994, p. 3) Consequently, institutions as well as organisations can have different units at different levels. This point is taken up in the analysis of Douglas and Wildavsky, where they argue that 'a successful bureaucratic hierarchy includes many interacting subunits'. (Douglas & Wildavsky 1982, p. 91) According to them the reason for sustaining a large organisation is to address a diversification of goals, to stop subunits from breaking away.

> Consequently the strategy emerges of compromise, of not pressing problems to the point of defining a simple overriding objective, but of creating a complex and obscure tradition in which subunits can find its place. (Douglas & Wildavsky 1982, p. 91)

The multiplicity and vagueness of goals has three advantages firstly it is 'easier to satisfy different elements of the firm', secondly, in retrospect 'to rationalise whichever ones happen to be accomplished' and thirdly 'their vagueness facilitates agreement on changing direction without appearing to go back on commitment.' (Douglas & Wildavsky 1982, p. 93) This approach can be linked to Ortner's understanding of the individual being loosely connected to structures, which could be contradictory in nature. (see above)

2.4 The outcome of the interaction of structures and actors

The debate of what came first or what determines the other, either the actor the structure or vice versa is not only very old, but also unsolvable. A combination of actor and structure is summarised by Hajer, 'social action originates in human agency of clever, creative human beings but in a context of social structures of various sorts that both enable and constrain their agency.' (Hajer 1995, p. 58) However, it would be difficult to identify the boundary between actor and structure, which seems to be 'not only highly fluid but also highly problematic'. (Meyer, Boli & Thomas 1994, p. 15) Both actor and structure are influencing each other; therefore an analysis would have to be actor and structure specific in a time and a spatial framework.

Debate arises again when it comes to the creation of institutions and the management of common goods. According to Olson, individuals would not co-operate if they have the same material interest. 'Unless the number of individuals is quite small, or unless there is coercion or some other special device to make individuals act in their common interest, rational, self interested individuals will not act to achieve their common interest'. (Olson 1971, p. 2) (coercion could again imply a form of inequality and a hierarchical structure) According to Olson, the self-interested individual tries to free ride and to gain benefits without contributing. 'Whenever one person cannot be excluded from the benefits that others provide, each person is motivated not to contribute to the joint effort, but to free ride on the effort of others.' (Olson 1971, p. 2) This approach is also reflected in game-theory and in Hardin's (1968) 'tragedy of the commons'.

However, new institutional approaches oppose these frameworks. The literature on common property resources points out, that individuals are able to work together for a common benefit.

> In situations of uncertain resource availability, common property regime members prefer to trade off some of the individual benefit generated by a system of private-use rights, for the collective assurance that the resource will be used equitably and as a consequence, sustainably. (Gibbs & Bromley 1989, p. 26)

The chapter on the local level will show that in the case of the water user associations in Karakalpakstan their argument is not valid. Hobley uses Bromley's argument but puts it into a wider structural framework. She argues, 'contrary to received wisdom, it is likely that as population increases common property regimes will become more desirable rather than less in those areas where prevailing cultural values support co-operation as a conflict solving device'. (Hobley, 1996, p. 2) The case study of water user associations in Karakalpakstan confirm Hobley's reasoning. However, prevailing cultural values do not necessarily imply equitable use of a resource. Hence, it will be shown that in Karakalpakstan there is no open conflict on the local level, because the cultural values do not allow to go against the former state and collective farm managers. Hobley's reasoning is contested by Alchian and Demsetz who argue 'The instability inherent in a communal right system will become especially acute when changes in technology or demands make the resource which is owned communally more valuable than it has been'. (Alchian & Demsetz 1973, p. 24) Even though Hobley uses the theory in a wider perspective, she is also limiting it. According to her, co-operation is only possible where old cultural values 'prevail'. However, the case study shows that co-operation does not imply equitable management over resources and therefore does not imply a just or fair system. Less culturally determined is Rawls. His theory of the just society defends Bromley's assumption. Rawls argues that 'the moral (just, co-operative) behaviour of one individual leads to increasing moral behaviour by others, reinforcing the co-operative behaviour of the first and encouraging still more'. (Mueller 1989, p. 15) The evidence of local management of common property resources indicates that it is possible to overcome the problem of free riding. The debate of creating institutions is directly interwoven with the debate on institutional change. Rawls's reasoning, is more relevant for the case of water user associations in Uzbekistan, because even though old norms are prevailing these norms are more a hindrance to equitable water user associations.

3 Institutional change

According to Hobley, 'institutional change focuses on the rules and processes that govern relationships between organisations and the public, and between different organisations. (North 1990, Ostrom 1990) It refers to changes in the architecture and relationships between agencies and organisations'. (Hobley 2000, p. 15) However, her approach does not consider norms and values. Hence a definition of institutional change has to incorporate changes in value systems, in informal and formal norms and rules of behaviour between agents, and secondly between agents and organisations.

3.1 All institutions are equal, some institutions are more equal than others

Different approaches seem to agree that institutions have different levels of manifestation and that some are easier to change then others. For example Hobsbawm in his analysis on the invention of tradition distinguishes between traditions and customs, while 'the object and characteristics of "traditions" [...] is invariance', customs do not 'preclude innovation and change up to a point' (Hobsbawm 1983, p. 2)

Alchian and Demsetz in their analysis of property rights argue 'the most important effect of alterations in institutional arrangements may well be the impact of such reorganisations on the cost of transacting'. (Alchian & Demsetz 1973, p. 22) Taking the approach of transaction costs, Binswanger argues that institutional change is itself a resource using activity. (Binswanger & Ruttan 1978, p. 334) Therefore the transaction costs of change have to be considered as a variable to determine the manifestation of the institution. The Coase-theory argues that institutions will be 'altered when the benefits from change will exceed the costs'. (Feeny 1988, p. 162) However, the context has to be considered. The costs and benefits for whom, for an individual or the whole community? This becomes very relevant by analysing two different administrative levels and the actors of the levels. It will be shown that the short term costs of the former state and collective farm manager is very high. The short term costs could imply no water for irrigation or a reduced amount. On the other hand the costs of the province manager for water distribution are very low. Hence, in his case the benefits are exceeding the costs.

Ostrom argues that 'rule changes may result from self conscious choice or may evolve over time as people develop shared understandings of what actions or outcomes may, must, or must not be done in particular situations. (Ostrom, 1994, p. 77) In the approaches of North and of Ostrom, it is pointed out that there are different levels of rules ordered in a hierarchical system, 'where each level is more costly to change than the previous one'. (North, 1990, p. 83) Ostrom distinguishes between three different levels of institutional rules:

operational rules, collective-choice rules and constitutional-choice rules. (Ostrom, 1994, p. 46) The author, Wegerich (2002), utilises Hajer and Bourdieu to effect and reasons that Ostrom's distinction of three levels is not sufficient enough. He argues that institutional change takes also place within discourse and within habits. (see below for Hajer's reasoning) The costs of change are dependent on the set of rules, and are the highest for alterations of constitutional choice rules. Ostrom argues, 'changes in deeper-level rules usually are more difficult and more costly to accomplish, thus increasing the stability of mutual expectations among individual interacting according to a set of rules'. (Ostrom 1990, p. 52) Hobley in her analysis of different institutional levels and change distinguishes also between three levels.

> There are three levels contained within the term 'institutional change', which have different implications in terms of the degree of change to be promoted and the means by which support is provided. 1. Institutional change (architecture: agencies and relationship) 2. Organisational change (capacity building, repositioning within agencies) 3. Process change (reengineering). (Hobley 2000, p. 15)

The discussion on different manifestations of institutions, indicates that the transaction costs for change vary. According to the theory, the costs for change are positively related with the level of institution. However, the case study will indicate that Ostrom's theory is not necessarily true, when the institutions are supply induced. The top-down introduction of water user associations will show that the operational rules do not necessarily change, even though the constitutional rules have already changed. Hence, the case study contradicts Ostrom's reasoning. An analysis for the reasons of change might help to determine the costs of change.

3.2 *Reasons for change*

Sahlins and Ortner argue in their anthropological analysis of society, that structures are constantly reproduced. However, established formal and informal institutions may change over time. So what determines during the process of structural reproduction which structures get altered? The reasons for change can be explained through the Rawlsian veil of ignorance, which is determined as 'an artifice of extreme information incompleteness'. (Shepsle 1989, p. 139) According to this approach, the individual or the community are only partially informed about the 'world' and the relations of subjects and objects in the world to each other. Hence, not all parameters are known. Individuals and communities establish institutions with their incomplete information. New information about the objects and subjects in the world is changing the definition of the position of the individual and the community. Taking into consideration that there are different structures, which colour the interpretation of knowledge and information, new information might trigger different responses. It might trigger new self-definition, communication, practices or technical changes. These in turn could lead to either new institutions or manifestations of the old formal and informal institutions. However, the focus on new information seems to be incomplete, new information is not the only cause of change. The individual has a variety of choices to respond to one structure, or (in the case of Ortner's argument different structural sets are contradicting each other) to choose different structures, decisions for possible responses may change the settings and can lead to new self-definitions of the actor, or to new definitions of the social environment about the actor. Again the new self-definition may lead to new practices, which might change informal and formal structures. Hence, structural reproduction plus 'an irreduceable present' might lead to changes. Furthermore, changes are not always intended. Therefore the reproduction of structure can have some additions, which are based on mistakes, which occurred during the process of reproduction. However, it is questionable whether new information, choices and mistakes lead to shifts in structures or only to a reproduction of the old structure in disguise.

According to the above approaches, every institution is vulnerable to influences and hence to change. The situation of the institution might be determined through inner and outer influences, which encourage changes, either in information or in the behaviour of individuals. According to Shepsle, institutional change occurs when a sufficient number of individuals are able to disrupt the institution. (Compare Shepsle 1989, p. 142) Shepsle's assumption links the number of individuals to institutional change. However, as the focus on different stakeholder groups indicates, the number of individuals is not the main determinant of change. It seems that the power position of the individual is important. Powerful individuals unsatisfied with the current situation can dominate the process of institutional change. Here powerful is relative, hence there might be more than one powerful actor in every setting, which could be in competition or in agreement over the changes The focus on power in relation to institutional change allows a combination of demand and supply approaches to induced institutional change. Hence, the focus on the role of power structures and power relations is a key to analyse institutional change. (see the debate on elites) This reasoning has a very important application for the case study. While the focus on the local level (see chapter four) indicates that the power of the former state and collective farm manager is sufficient enough to block the introduction of equitable water user associations, the focus on the provincial level (see chapter five) shows that a more powerful actor can implement changes.

3.3 Introduction to the debate on institutional change

In the New Institutional Economics (NIE) literature two different approaches to institutional change are used: demand and supply induced change. Hechter calls them co-operative and imposed institutional change. (Hechter 1990, p. 14) Anyhow, the terms of demand

and supply in case of institutional change can be interpreted in a developmental perspective leading to the paradigms of bottom-up and top-down.

Keohane in his analysis of international regimes combines the demand and supply side explanations. He argues, that the demanders and suppliers of regimes are likely to be the same actor or elite. This implies that the demanders of an institution are always able to introduce an institution and to make it sustainable, and this reasoning is very questionable. However, Koehane reasons more convincing at a different point that, 'factors affecting the demand for international regimes are likely simultaneously to affect their supply as well.' (Keohane 1983, p. 142) However, while international regimes are already on the highest level, it might be possible to reason that this would be different for the local, district, provincial or national level. Hence, here the demanders of change are not necessary the suppliers of the changes.

NIE scholars have argued in their earlier work that the main reason for change is minimisation of transaction costs and efficiency. 'It is as if it were being assumed that institutional innovation took place in a perfectly competitive political arena guaranteeing that only efficiency-improving innovations would be selected.' (Feeny 1988, p. 165) However, as mentioned above North, the main advocate of demand induced change, had to abandon the claim that institutional change is an attempt to increase efficiency. North argued that inefficient institutions stay unchanged if they have the necessary support. De Janry argues that dysfunctional institutions stay in place either because of 'path dependency, the Prisoner Dilemma or imperfect information about future gains'. (de Janry 1993, p. 567) He reasons

> in the first explanation, high sunken costs induced by past institutions make transition to what could be a superior institutions socially unprofitable. In the second explanation, without cooperation, the cost of breaking an existing rule may be too high on individuals, and it is thus individually rational for socially sub-optimum institutions such as castes to remain existence. (de Janry 1993, p. 567)

A different perspective is taken by Garston, who argues still in line with North, 'institutions may persist in behaviour and structure may be maintained, when neither serves the current interests of participants'. (Garston 1993, p. 17)[7]

3.3.1 Demand induced change – bottom-up change

North, taking a demand management approach to institutional change, argues,

> the agent of change is the individual entrepreneur responding to the incentives embodied in the institutional framework. [...] Change typically consists of marginal adjustments to the complex of rules, norms, and enforcement that constitute the institutional framework. (North, 1990, p. 83)

For North the most important source of institutional change is the change of relative prices. On the other hand Feeny argues, 'the basic source of the demand for changes in institutional arrangements is the recognition that existing arrangements leave potential gains un-captured.' (Feeny 1988, p. 176) Alchian and Demsetz looking at shifts from common property regimes to private property regimes argue that changes in technology and demands make the resource more valuable. (Alchian & Demsetz 1973, p. 24) Hence, changes in prices for the resource may lead to changes in institutions. These approaches are not

---

7 Garston's reasoning is similar to the distinction of Hobsbawm between traditions and customs. Garston states 'the interplay between the two aspects (ceremony and function) of organisations is seen as highly related to changes in the technology employed, since what is functional once may become ceremonial when methods of production change'. (Garston 1993, p. 17) However, while Garston's approach only relates to technological changes and production, Hobsbawm's approach focuses on customs, which become fixed in time, because of political decision to manifest status and values. Keohane validates this reasoning and argues 'considerations of power and status may be more important than the functions performed by the institutions' (Keohane 1994, p. 53)

necessarily contradictory. Taken into account that the veil of ignorance is slowly lifted, new information may lead to the recognition that existing arrangements leave gains uncaptured as well as new information may change the relative prices for a resource.[8]

The introduction of the parameter 'information' links technological change and institutional changes. NIE literature points to the interdependence between technological and institutional change, whereby both influence each other and can induce change of the other. Hechter links the formation of institutions with changing environmental and demographic conditions.[9] Hechter's conclusion can be utilised for the debate on institutional change. Institutions in place may adapt to increased or new demands, which are caused by environmental and demographic changes. Ohlsson (1999) and Homer-Dixon (1995) link social, adaptive capacity and environmental and demographic changes. (water scarcity, due to environmental, demographic or developmental factors might induce shifts within society, when the adaptive capacity of society is high, see below on adaptive capacity) Their concept can be applied for the theory on institutional change as well.

Hechter, looking at the formation of institutions argues, 'demand alone is insufficient to produce co-operative institutions', and he reasons 'individuals must be highly visible to one another in order to reduce the severity of the free rider and assurance problem'. (Hechter 1990, p. 19) His statement is useful for the debate on institutional change. Because it shows that demand alone is insufficient to induce changes. Demand has to be conjunctive with a potential control capacity. Keohane reasons that the formation of international regimes is determined by 'relationships of power and dependence in world politics' and 'constraints are dictated not only by

8 The theory of demand-induced changes is a bottom-up approach. It implies that small changes at the bottom influence the whole structure.

9 Hechter reasons 'if institutions emerge as a result of the demand for joint goods, then shifts in a variety of environmental and demographic conditions will heighten demand for certain kinds of joint goods and favour the emergence of institutions supplying these goods'. (Hechter 1990, p. 18)

environmental factors but also by powerful actors' (Keohane 1983, p. 146) (see below debate on elites) In the context of institutional change it means that the success of demand-induced changes is based on power.

However, North's approach is very limited, because he refers to institutions, focusing mainly on property rights, leaving other approaches to institutional change unconsidered. However, Keohane argues, 'understanding how people think about institutional norms and rules, and the discourse they engage in, is as important in evaluating the significance of these norms as measuring the behaviour that changes in response to their innovation'. (Keohane 1994, p. 46) Hajer, using an argumentative approach (communication science) to institutional change argues that the rules-in-use change when the implicit meaning of terms within communication changes.

> Rules, distinction, or legitimate modes of expression, only have meaning to the extent that they are taken up. It implies that the rules and conventions that constitute the social order have to be constantly reproduced and reconfirmed in actual speech situations, whether in documents or debates. (Hajer 1995, p. 55)

Hence, institutional change takes place within discourse.[10]

3.3.2 Supply induced changes: change from above and from outside

The demand approach to institutional change does not take into consideration that change can be induced from above. Binswanger refers to supply-side induced change, which can either be induced from above, within the institution, or change can also be induced through outsiders. He argues, 'institutional change may occur as a result of advances in the supply of knowledge about social and

10 Or utilising Luhmann's approach 'communication between human beings is what makes society a reality'. (Papadakis 1996, p. 33) The manifestation of changes such as law, are only the last evidence of change.

economic behaviour, organisation and change'. (Binswanger 1978, p. 334) He points to similarities between institutional and technical change. In both cases change is possible through advances in knowledge in specific disciplines. In the case of institutions, change could be induced by shifts in knowledge in the social sciences and their related professions. (Binswanger 1978, p. 338) Binswanger's approach can be utilised at different levels of institutions. Hence outside knowledge does not necessarily change the whole institutions but only parts, which are directly influenced by new knowledge. However, for these parts the knowledge is provided from above and changes are implemented in a top-down approach. Binswanger's reasoning can be related to the case study. A supply induced change occurred when it was ordered from the national level to abandon state and collective farms and to create farm organisations and water user associations on their territory.

Downs' assumption of the rational and selfish individual has consequences for institutional change. It implies that the politician's individual return will be different to the social return and that the 'institutional innovation will not be supplied at the social optimum level'. (Feeny 1988, p. 169) This reasoning becomes important in the implementation phase of land reforms and the establishment of water user associations. The chapter on the local level will show how not politicians but former stakeholders implementing the changes will supply the reforms, but not at the 'social optimum level'. In addition, arguing from Downs' perspective, one could reason that governments do not have a developmental goal and only organised pressure groups determine political goals and influence politicians. This is possible because the main goal of the politician is to stay in power. Therefore, they act according to the interest of powerful pressure groups, who could influence their re-election. This reasoning confirms the analysis of Olsen on collective action.

> Both theories (public choice and collective action) suggest that the markets for economic and political resources wherein the demand for institutional change is expressed are so imperfect as to create fundamental biases in the

> direction of institutional change. These biases generally favour organised producers over consumers. (Binswanger 1978, p. 351)

Or as will be shown in the case study elites over the general population. However, Khan argues against the purely rational and selfish motivation of governments. Indeed as argued in the analysis of the actor oriented approach the introduction of relevance contradicts Feeny, hence institutional innovation can be supplied at the social optimum level, but only if the innovation becomes relevant for the power stakeholders. Binswanger's reasoning is confirmed by the non-optimal introduction of water user associations on the local level. However, the shift of the argument is realised in chapter five on the provincial level, here the introduction of functioning water user associations becomes relevant to the provincial manager of water distribution.

However, Binswanger is himself critical of the approach of supply induced institutional change, as mentioned earlier he sees similarities between institution and technology. He argues 'inappropriate institutional transfer results in biases in the supply of institutional change that are similar to the biases that inappropriate technology transfer introduces into the supply of technical change.' (Binswanger 1978, p. 353) He continues, 'institutional innovations are probably even more location-specific than are technical innovations.' (Binswanger 1978, p. 355) His reasoning indicates that he perceives established institutions to be less flexible than technology. Hence, supply induced institutions have to fit certain criteria, to be successfully implemented.[11] Gibbs and Bromley argue that the institutional environment and technology have to complement each other to function sustainably and efficiently. (Gibbs and Bromley 1989, p. 23) However, their reasoning can be interpreted in the way that institutions are only supplementing technology, and that it would be enough for induced institutions to support the existing technology

11 However, currently outsiders, such as the World Bank, are promoting almost identical reforms, throughout the developing world quite independently of national context. (Personal communication D.J.W. Berkoff 27.06.2003)

to become sustainable. According to their reasoning technology and institutions are not equal factors. A different approach is taken by Berkes, arguing that institutions 'cannot be created anew', and that 'local people cannot be divorced from the social structures of which they are a part'. (Berkes 1989, p. 13) Keohane confirms this reasoning and argues 'just as the actors in world politics are constrained by existing institutions so are institutions and prospects for institutional change, constrained by the practices taken for granted by their members.' (Keohane 1994, p. 49) This would imply that new institutions have to sustain the old institutions to be successfully implemented and to be sustainable. Feeny argues from the point of the old institutions, for him it is evident that 'existing institutional arrangements affect the supply of institutional change'. (Feeny 1988, p. 188) This confirms Douglas' and Bourdieu's reasoning about institutions and their effects on institutional change. Feeny's statement raises the issue of institutional path dependency. The concept of path dependency is accepted by economic historians as well as political scientists. (Papadakis 1996, p. 27) However, the concept of path-dependency does not incorporate wider political changes in power structures and shifts in hegemony. (Compare Young 1989) Hence, the background chapter will show, how the Soviets tried to erase the structure of local water management and tried to create a new structure on the ruins of the old one. Only tried, it is debatable whether they succeeded, and if they did whether the created structure is what they wanted to create or is only a partial reproduction of the old structure. This question will be addressed for the new reforms in the conclusion. The simplified version of the concept is given by Papadakis who states 'where you get to depends on where you are coming from, and some destinations you simply cannot get to from here'. (Papadakis 1996, p. 27)

The concepts of 'emblematic events' (Hajer 1996) and 'windows of opportunity' (Kingdon 1984) seem to contradict the above reasoning of path-dependency, because they imply that institutions can escape the 'inescapable' past, and are solely determined by an 'irreducible present'. However, structuralism and institutional theory deliver explanations for these kinds of concepts. Sahlins' theory takes away the borders between structure and event. Every event influences the individual, which reproduces the structure (see Sahlins approach on institutions), consequently a structure should change constantly with the occurrence of new events. However, the event is perceived within the framework of the structure, 'it is a relation between a certain happening and a given symbolic system' (Sahlins 1987, p. 153). Therefore the event cannot be understood without the structure, and it is impossible to see the event opposed to a structure. Papadakis confirms Sahlins' reasoning, and argues that new concepts only become significant 'if the opportunity arises', and 'the opportunities are of course determined [...] by institutions and by the prevailing rules of the game.' (Papadakis 1996, p. 25) Hence, the institutions influence the perception of occurring events. Events can be recognised or stay invisible, depending on the readiness of the institutions in place to emphasise them or to de-emphasise them. The introduction of farm organisations and water user associations on the territories of the former state and collective farms could have been a window of opportunity for the newly independent farmers to break with the top-down institutions and to use the new institutions to demand changes of water management or training. However, the focus on the local level will show that the window of opportunity was de-emphasised as real change and that the institutions stayed the same.

A less deterministic and more actor orientated approach to institutional change is taken by de Janvry. He reasons that the key to understanding the influence of the institutional environment 'is to identify which agents have the incentive and the capacity to implement institutional change and how well these agents are defined.' (de. Janvry 1993, p. 567) Feeny adds to such capacity the willingness needed to implement change, he argues, 'the supply of institutional change depends on the capability and willingness of the political order to provide new arrangements.' (Feeny 1988, p. 182) Papadakis, analysing the political order and its influence on institutional change, concludes, 'established political organisations are usually better placed than new ones both to take advantage of and to influence the institutional framework'. (Papadakis 1996, p. 6) On the other hand, de Janvry reasons that 'policy makers do not have an objective of their own, and are simply motivated by staying in office, doing for that purpose whatever it takes'. (de Janvry 1993, p. 569) His reasoning confirms Downs' belief in the rational actor in politics and bureaucracies. De Janvry's argument does not contradict Feeny, however, because he argues

> there are a number of important factors that affect the capability and willingness of the political order to provide new arrangements. These factors include the cost of implementing the new arrangements, the normative behavioural code, conventional wisdom, and the expected net benefits to powerful elite decision makers who exercise positions of dominance. (Feeny 1988, p. 183)

Feeny tries to combine actor and structure approaches. The main actors seem to be elite groups influencing decision-making or making the decision themselves. Feeny states more blandly the 'political and economic costs and benefits to the ruling elites are a key to explaining the nature and scope of change'. (Feeny 1988, p. 168) However, missing in Feeny's approach is the consideration that elites operate on different spatial and social levels with different functions. Elites are on the national, regional and local level and within their sphere of influence they might support or resist changes of institutions at multiple levels. Also the ability of elite groups to operate at different levels can make them more effective at a specific level compared with a group acting at a single level.

Feeny states that the costs of implementing supply-induced changes are significant. He links supply-induced changes to the transaction costs of change. While Binswanger argues that supply induced changes are cost minimising, 'the possibility of borrowing institutional innovations [...] reduces the costs of institutional change' (Binswanger 1978, p. 352) Allan's analysis of innovation as the sequence 'knowing, wanting, having, operating and efficiently operating' (Allan 2001, and personal communication) indicates that knowledge does not necessarily imply a reduction of transaction costs.

On the contrary it could be argued that it is possible that knowledge creates opposition and leads to an increase of transactions costs. In terms of the case study, it would be important to ask who knows and who wants the institutional change. In addition, as Allan's approach indicates it is necessary to differentiate between operating and efficient operating, hence further, and often very substantial transaction costs arise. Shepsle validates Allan, and reasons that changing institutions impose transaction costs. He specifies the kind of transaction costs, which are bound to rise. 'The transaction costs include not only those devoted to decision making, but also those required to enforce the procedure of the new regime and for individuals to adapt to the new procedures.' (Shepsle 1989, p. 144) 'The costs of police forces, law courts and penal institutions are evidence of the phenomenon' (Allan 2001, personal communication) In the literature this problem is referred to as 'second order public good problem'. The first order problem is to establish the institution. Coleman argues that sanctioning is a 'second order public good problem', because the benefits of sanctioning is not as high as the costs for action. (Coleman, p. 52) Ostrom argues that not only sanctioning but already monitoring could be defined as a 'second order public good problem'. (Ostrom, 94, p. 48)

The conceptual division between the demand and supply versions of institutional change, suggests that there is either one or the other. Mary Douglas in her anthropological analysis of society takes both processes into consideration. Douglas divides society with its institutions into centre and border. In her theory the centre is constituted of hierarchical and individually organised institutions; the border of voluntary organised institutions. She argues that the 'centre is too constricted in its casing of institutional habits' and claims that 'no change ever comes from the centre, all innovation comes from without'. (Douglas and Wildavsky 1982, p. 189) 'The outsider sees more clearly and renewal comes form the margins of society.' (Douglas & Wildavsky 1982, p. 189) Changes suggested from the centre focus not on real alternatives but on suggestions, which are 'best known and closest to existing programs'. (Douglas & Wildavsky

1982, p. 93) However, it is problematic to differentiate between the centre and the periphery of institutions, and marginal and real changes. The boundaries between the different terms are weak. Douglas' concept of only marginal changes from within, is in agreement with Bourdieu's approach of habits and change. Her distinction between marginal and real changes, does enhance the NIE debate, which does not give sufficient attention to change. In Douglas' approach, 'real' changes alter the institutions while 'marginal' changes reconfirm the institutional structure, without changing it. A problem with the Douglas approach is, whether it can be determined if the agent of change is an outsider acting within the institution. Other questions, which stay unanswered, are: What qualifies an individual to be an outsider to an institution and to what institution is the individual an outsider? (For example 'epistemic communities' see Haas 1994, p. 128) (Haas gives the example of engineers coming into positions in Parliament and changing the discourse and the perception of water.) Even though, Douglas' approach seems to be helpful in determining the nature of change and the analysis of such change, it also complicates the analysis.

Overall, it seems that institutions change, when benefits to the stakeholders within the institution are high, and when the alteration does not threaten their interest. The observations of the local level in Karakalpakstan and Khorezm show that the introduction of functioning water user associations threatened the interests of the former state and collective farm managers. Hence, water user associations were officially implemented, but were not functioning. However, it is necessary to distinguish between different stakeholder groups at different institutional levels and in different space/time circumstances. The distinction is important because a supply side institutional change from the top level can cause resistance or only partial adaptation on the lower levels. This could arise because of regional or local level resistance to change from above. On the other hand, local changes, which threaten higher-level stakeholders, may face resistance from above. Water management and water management institutions beyond the customary level are hierarchical.

Social and institutional theory is very helpful in linking actors, structures and interacting processes, which enable and shape institutions and the pace of institutional reform.

4 Institutional adaptive capacity: social capital and the role of elites

4.1 Influences on institutional change: collective action and elites

As the discussion about institutional change indicates, the power of interest groups is one key to understanding the dynamics within institutions. Power-relations and hierarchies are very significant in stakeholders' interaction. Olson argues that inequalities are beneficial in stimulating and sustaining collective action. Das Gupta on the other hand argues, that 'strong socio-economic hierarchies seriously undermine the potential for collective action'. (Das Gupta 2001, p. 1)

Baland and Platteau claim that 'There is abundant evidence to support the hypothesis that the costs of initiating collective action are largely borne by the economic elite.' (Baland & Platteau 1999, p. 780) They argue that for example the rural co-operatives in the Netherlands were often created by groups of influential, better-off farmers. According to them, these farmers could bear the cost of initiating collective action. In addition Wade argues that the effectiveness of a local irrigation council 'depends on its councillors all having a substantial private interest in seeing that it works, and that interest is greater a larger a person's landholding' (Wade, 1987, p. 230). Furthermore, big landowners' claims on common pool resources 'are sufficiently large for some of them to be motivated to pay a major share of the organisational costs' (Wade, 1988, p. 190). Baland & Platteau argue that on the other hand poorer farm owners do not initiate collective action. Reasoning that each of them internalises too small a share of the benefits resulting from well-managed common pool resources.

However, the wealth endowment has not only positive effects. Mosse argues that some individuals have the capital to 'deviate from the rule without attracting public notice or sanction'. (Mosse 1997, p. 481) He argues that the poor, lower caste or women farmers do not have this possibility. Baland & Platteau make a similar observation. 'Wealthier users can not only refrain from participating in resource-preserving collective actions, but they may also attempt to under-mine such actions in order to further their own private interests.' (Baland & Platteau 1999, p. 782) Shan-mugaratnam links participation of the wealthy with exit opportunities and political capital, arguing,

> Absentee herd owners favour open access rangelands so that their herds can graze anywhere. They may even use their political influence to prevent pastoral associations receiving legally defensible land rights. (Shanmugaratnam et al., 1992, p. 20)

Even though in the presented debate on collective action it is focused on elites defined by wealth, in the case of Uzbekistan the local elites are not defined by wealth but by the position within the institutions. Their power position enables them to act similar to elites defined by wealth.

Balland & Platteau analyse social heterogeneity and its effects on collective action. They argue that 'poor stakeholders might have shorter time horizon and therefore have to choose strategies which yield more immediate results, and disregard longer-term considerations in resource conservation.' (Baland & Platteau 1999, p. 774) They reason that depending on the level of wealth poorer users may not participate because collective action violates their survival constraint. This would imply that the better off farmers would have the main burden of contribution. Furthermore, it implies that the poor would infringe the management rules of the community to a larger extent then the rich. However, Robbins (2000) analysing corruption in forest management in India shows that the local elite has easier access to common pool resources. Robbins argues that the higher caste elites occupy important positions in government, policing and forestry. The social capital of the local elite reduces the access costs to the forest resources. Lower cast elite which do not have connections to position holders have to pay higher 'entrance fees' in the form of preliminary bribes, the higher entrance excludes poorer households as well. This becomes relevant in terms of access to irrigation water on the former state and collective farm. Because of their power position the old stakeholders have easier access to the water resources, and for new land owners without social networks it would be more difficult and more costly to get water.

According to collective action literature, it seems that 'commons work best when a recognised and legitimate leadership can regulate use' of the common pool resource. (Ensminger & Knight 1997, p. 8) Blower and Leroy confirm this view in their analysis of community collective action against 'locally unwanted land users'. Here, the role of the local elite determines the success of local initiatives. (Blower & Leroy 1994, p. 207) Mosse argues that 'leaders with wide-ranging influence are able to act as 'water brokers', arranging the purchase of water and its delivery or sale to villagers en route.' (Mosse 1997, p. 479) While in Mosse's example kin and caste links are determining leadership, Wade argues that big landowners can initiate collective action and take the leadership. (Wade 1987, p. 230) However, as already argued above wealthier users might also undermine collective action. This could also be the case for old leaders. Exit opportunities can increase the bargaining power of middle-income farmers, which might lead to an increase in political capital of these groups. Even though, in the literature exit-opportunities are only defined as possibilities for alternative income, it would be possible to define exit opportunities in terms of having alternative social networks, which would enable new actors to create a balance against old established actors. In the case of the water user associations in Karakalpakstan and Khorezm the land reforms brought actors with networks to higher levels down to the local level. These new actors have the possibility to challenge the old local leaders, on the basis of their networks. Hence, old and new leaders might be in competition over power. Hence, leadership does not necessarily imply collective action over common pool resources. The argument is validated in the case of the water user associations in Karakalpakstan and Khorezm, even though there are established leaders, it does not imply that collective action takes place. On the contrary, the old leaders can block the establishment of collective action organisations.

Blau and Scott (1969) analysing common pool resource management structures, use the term mutual-benefit organisations. According to them these organisations have the problems of maintaining democratic processes. Apathy of the members is common as well as the development of oligarchic control through active minorities. They argue that sometimes within mutual-benefit organisations the democratic controls are sacrificed for the accomplishments of objectives. In addition, studies indicate that 'persons of higher socio-economic status tend to belong to more associations and to participate more actively in them than persons of lower status'. (Blau & Scott 1969, p. 105) The combination of the two factors shows that even though a democratic organisation for common pool resource management is set up, it does not imply that the original system continues.

4.2 Influences on institutional change: adaptive capacity

The influence of hierarchies on institutions is prominent in the literature on adaptive capacity. In this review the concepts of social ingenuity (Homer-Dixon, 1995), social adaptive capacity (Ohlsson, 1999) and social capital (World Bank late 1990s) are used interchangeably. Giddens' concept of reflexivity aims to explain the same phenomena. All four approaches are based on the same concept; namely that a high level of the social good, that enables change and reduces transaction costs, improves the capacity to deal with crises and uncertainty, whether the crisis is political, economic or environmental. As Homer-Dixon formulates 'Social ingenuity is key to the creation, reform and maintenance of public and semi-public goods such as markets, funding agencies, educational and research organisations and effective government. (Homer-Dixon 1995, p. 4) Giddens argues that reflexive societies can adapt to changing conditions, the higher the level of adaptability the more reflexive the society. Hence, a highly flexible society has more chance of survival. Furthermore, it is argued that a high level of this social good will enable individuals and groups to change institutions. Homer-Dixon argues that 'technological ideas are not the only productive ideas; just as important are ideas about social organisation, especially about reforming and building institutions' (Homer-Dixon 1995, p. 2) Edwards agrees and argues,

> the value of civil society lies precisely in its transformative potential: the ability of citizens to lever fundamental changes in the market economy, and in the structures of politics and social relations that underpin it. (Edwards 2001, p. 4)

Where does this social good exist and develop? Homer-Dixon argues, 'the generation and dissemination of productive ideas is endogenous not just to the economic system but also to the broader social system that includes a society's politics and culture.' (Homer-Dixon 1995, p. 2) The research of the World Bank points to the fundamental and enabling structures of society. The Bank's research indicates that 'strong social networks and a dense web of civic associations are likely to facilitate certain aspects of economic performance.' (Edwards 2001, p. 4) However, not every network is positive, as Woolcock points out social capital can also be associated with costs, 'such as cults, group loyalties, destructive acts of hate groups, drug cartels, terrorist organisations, nepotism laws' (Woolcock 1999, p. 7) The focus of those wanting to enhance economically beneficial social capital would have to pick on 'winners' which promote social behaviour which is wanted by powerful in- and outsiders. (The question arises what is wanted? Ideologies, contend and dominate, can be religious, believe systems, economic or political. An example would be the new forms of colonisation, Northern emphasis on democracy, feminism and environment dominating the NGO agenda in the South) Edwards who critically analyses economic development strategies argues 'the easiest areas to influence in the short run – like the number of NGOs and other civic organisations – may not be especially important in the grand scheme of things.' (Edwards 2001, p. 8) He points out that the 'social enrichment is a matter of behaving differently toward each other, not just adding on more social contacts.' (Edwards 2001, p. 5) He concludes that, 'the most important areas of social capital are also the most contingent and the most difficult to engineer – attributes of trust, tolerance and non-discrimination.' (Edwards 2001, p. 8) All these are norms, informal and non-legislated institutions.

However, Edwards points out that the norms and behaviour of a community's 'social capital can disempower poor people where the wider political setting is not supportive, and political and economic inequalities are not addressed.' (Edwards 2001, p. 5) This is confirmed by Putnam who argues that, 'social inequalities may be embedded in social capital. Norms and networks that serve some groups may obstruct others, particularly if the norms are discriminatory or the networks socially segregated.' (Putnam 1993, p. 8) Woolcock points out that the presence of 'equality', 'liberty' and 'good governance' enhance social capital. (Woolcock 1999, p. 4–5) Putnam who researched communities in the North and South of Italy argues that in successful communities 'social and political networks are organised horizontally, not hierarchically'. (Putnam 1993, p. 2) He further argues that 'a society that relies on generalised reciprocity is more efficient than a distrustful society. Trust lubricates social life' (Putnam 1993, p. 3) On the other hand, if the society is already unequal, then trust might lead to further exploitation of the poor by the rich and influential. Hence notion and extent of trust are scale related, or as Edwards, argues 'generally, poor people do best when they are cautious reciprocators'. (Edwards 2001, p. 6) Overall the analysis shows that, inflexible hierarchies within society prevent the development of social capital, ingenuity or adaptive capacity. This observation of the local level confirms the statement. Here, the inflexible hierarchies prevent equal water distribution among the new farmers from occurring, instead the old hierarchies obstruct the adaptiveness of the local community.

4.3 Institutions and elites

The analysis of agricultural production and water management in Uzbekistan will show that the role of local elites is central in the old communities and the new institutions set up during the Soviet Union. The discussion about institutional change and adaptive capacity shows that, powerful stakeholders are the key to understanding the dynamics within society. Some stakeholders can be defined as elites within a society. Marcus connects elites with institutions and points out that elites are 'intimately associated with the working of institutions since their informal organisation evolves along with or in opposition to formal organisation in a complex society'. (Marcus 1983, p. 10) In early definitions of elites, they were understood as 'groups situated in a social system, in which they, as elites, dominate an institutional order', (Marcus 1983, p. 12) in modern societies elites are defined 'as creatures of institutions in which they have defined functions, offices, or controlling interests'. (Marcus 1983, p. 16)[12]

However, elite organisations differ from institutional structures, in the sense that elites 're-create a domain of personal relationships that extend across functional and official boundaries.' (Marcus 1983, p. 16) Elites interact with other elites in other institutions. They create networks, which are not necessarily functional for the formal institution nor are they linked with the formal institutional hierarchy. While this explains relationships between different

12 Modern societies are determined by six principles: industrialisation, centralisation, statism, secularisation, decrease of importance of religion, individuation. However, the distinction between modern and pre-modern is dangerous, it allows the rise of simplification of 'the other'.

institutions or different levels within the institutions, it does not fully explain how elites and institutions interact. Marcus argues,

> elite organisation accomplishes the reconstitutions of one kind of rules (bureaucratic codes and procedures that are publicly available) through the creation of other rules of a different type (tacit, operational rules) in order to make formal organisations serve the interests, however defined, of the members of the elite. (Marcus 1983, p. 42)

Hence through institutionalisation and institution building they are able to keep and enhance their role in the power structure. During the Soviet era agricultural production was institutionalised, the new institutions required specialists which managed the farms, the technology on the farms and the irrigation structure. The institutionalisation altered the perception of agriculture from pre-modern to modern.

In pre-modern societies elites kept their power through the patriarchal and clan systems, in modern societies elites 'can only offer expertise and professionalism as a legitimation for their positions'. (Marcus 1983, p. 52) But in modern societies legitimacy is not only achieved through technical knowledge, it can also be based on functions defined in law. Legal legitimacy 'has evolved as the ultimate, abstract source of authority in mass liberal society.' (Marcus 1983, p. 52) Marcus argues that the justification of the authority of the elites is based on law or by the

> claim to serve the institutions of society in disinterested and rational ways, a claim that is the best public legitimation of themselves that elites can offer in a world organised by rational and apparently democratic, processes. (Marcus 1983, p. 52)

In the literature review it has been argued that elites have positive as well as negative influences on institutional change. In Karkalpakstan and Khorezm the old elites have legitimacy because of their training and specialisation. The institutions allows the old elites to hide their personal agenda and to appear to be legitimately elected in their new 'participatory' organisations.

4.4 Elites and demand induced institutional change: bottom-up

While Marcus emphasises the power of elites based on technical knowledge, Das Gupta points out that in the absence of well-functioning market and non-market institutions elites monopolise 'sources of information and contacts', hence, elites can be the gate-keepers to the outside world. (Das Gupta 2001, p. 4) Their position allows them to offer security (in terms of finance, labour, natural resources) to their dependents, a patron-client relationship. Young argues that this relationship creates informal structures and habits, which become more manifest.

> There is no reason to assume that dominant actors must coerce subordinate actors continuously to ensure their conformity to the requirements of imposed institutional arrangements. Habits of obedience on the part of subordinate actors regularly arise over time. (Young 1989, p. 89)

Das Gupta confirms this and she states

> over time, social norms emerge from repeated games between contracting parties. As anthropologists and historians have observed, these agrarian relationships generate not only rules of behaviour but also crystallise a moral universe characterised by an unequal distribution of socio-political power. (Das Gupta 2001, p. 3)

Hence, the patron-client relationship shapes the mind-set of the clients, and this might 'constrain their future mobility' (Das Gupta 2001, p. 4) Conclusively, she argues that 'It is abundantly clear to them that their safest strategy is to maintain good relations with their patron and never to challenge him [the patron].' (Das Gupta 2001, p. 4) Das Gupta's reasoning becomes important in the case of farm organisations in Uzbekistan, because the former local stakeholders continue to be in power positions as managers of the new organisations. Even though the new organisations should be bottom-up the old habits of patron-client relationship influence the behaviour within the new organisation.

As already mentioned above, Blowers and Leroy (1994) argue that a local political elite is important in generating environmental movements against 'local unwanted land users'. They argue that the resistance against these users stays or falls according to the support of the elite. Blowers and Leroy see the local elite as initiators of and the backbone of, collective action. Missing in their approach is the link between institutional change and the role of elites in a broader framework. Even though Blower and Leroy state that elites are important for local environmental movements, they do not link motivational aspects of elites to the action taken by the elite. It seems to be evident that elites encourage changes, such as environmental action, if these changes do not change their position or enhance their status within their community. Furthermore, because of contingent events or influences, issues can become more relevant to elite groups which might trigger their willingness to support and to further enable changes from below.

4.5 Elites and supply-induced institutional change: top-down

Adams in his research on supply induced economic changes, argues that the ruling elite is typically susceptible to institutional reform when 'international financial institutions convince one key political group that their political longevity depends on pursuing reforms'. (Adams 2001, p. 14) In a broader framework it could be argued that either outside or inside epistemic communities convince the ruling group, that their security is threatened or could be enhanced (security is this case is defined in terms of physical security as well as reputation towards insiders as well as outsiders) by political change. Hence, the drive for reform is selfishly motivated. As argued above, the selfish motivation is not enough to explain wholly the actions of the actor or of groups. It is more viable to argue that an epistemic community can convince a ruling elite about the relevance of a certain issue. This issue can become relevant to the ruling group and can be the driving force for the reform process.

Adams differentiates between eight different stakeholder groups. He argues that different stakeholder groups at different levels resist changes from above. Each social group either 'gains or loses utility in the policy reform process'. (Adams 2001, p. 7) Their utility is defined through their sphere of influence. Das Gupta makes a contribution to explaining the process by stressing that patrons have 'to make sure that their underlings do not obtain opportunities for upward mobility'. (Das Gupta 2001, p. 4)

According to Adams, 'if a particular reform engenders enough opposition among social stake groups with large political weights, that reform will be delayed and/or not implemented'. (Adams 2001, p. 3) However, Adams looks only at the political weight of the elites in place and on formal resistance, which either leads to delay or non-implementation. Scott indicates however, that there are many forms of passive and active resistance. Resistance does not have to be formal, but can be informal and effectively manipulate the intended reforms. Wienicki (1996) shows that middle management and unions in socialist countries are able either to manipulate changes or to weaken the suggested changes, so that the likelihood of failure is increased. Wienicki's case study indicates that the formal political weights are not that important, but more the position of the patron and the informal influence connected to the position.

While initially a reform may start because of its relevance for the ruling elite, for the remaining elite groups these reforms might cause a loss of utility. Hence, according to the influence of elites on and for different spheres, it seems reasonable to assume that a lower level elite could try to secure their position against changes inflicted by the ruling elite or an external patron elite such as the World Bank or the European Union. The lower level elite would either try formally to delay or abort the reforms or informally to manipulate the induced changes to their own benefits. This might have consequences for the implementation of the reforms. The changes put into practice might not reflect the initial plans. Das Gupta points out that in such change processes 'the patrons expect to be the prime beneficiaries of any new opportunities made available in the outside world'. (Das Gupta 2001, p. 4)

The implementation of reforms on the local level is very relevant in the case study, because state and collective farms are abandoned and new institutions, such as the farm organisations and the water user associations, are introduced. Reforms could be interpreted either as neutral or as a positive or negative for the local elite. In the case of negative effects, the local elite can resist or manipulate the changes. Depending on the influence of the local elite on the implementation process, the reforms are delayed, abandoned or informally adjusted to their demands. The adjustments can either be officially within the explicit rules or unofficially through non-evident influence. The more influential or powerful the local elite is the higher is the possibility of resistance, adjustment and modification of the reform initiative. Again it seems that the power of stakeholders is a key factor.

4.6 *Conclusions on elites and institutional change*

The literature review indicates that, real 'institutional change is considerably more difficult in highly inegalitarian settings' (Das Gupta 2001, p. 21) Hence, power stakeholders or elites are influencing institutions and their adaptability to change. The influence of the elite is even higher, when the elite controls 'gates' to the outside world, hence when they restrict the mobility of individuals or the flow of information. This is confirmed by Conning, who argues that the influence of the elite on changes is higher the 'less mobilised (informed) voters are at the local level'. (Conning 2000, p. 20) In addition, the influence of elites on change is enhanced by the notion of habits. Adams states, confirming Das Gupta's argument on the relation between habits and informal relations, that, 'political longevity is assumed to directly and positively influence the utility of the ruling elite and urban politicians in some unspecific, multiplicative manner'. (Adams 2001, p. 7) According to the findings the more in terms of length and control an elite group is established the more positive or negative influence has the elite on the process of the induction of institutional change. Hence, if the former state and collective farm managers were for a long period in their position, their influence is more dominating and it would be more difficult to change the power structure.

5 Risk, elites and institutional change: relevance of risk theory

Adams makes the connection between elites and their behaviour during risk situations, he argues

> Since the ruling elite of any country are uncertain about the benefits that either they or the members of their society will enjoy if a particular reform is adopted, the first principle – self interest – becomes the determining factor. (Adams 2001, p. 14)

The following analysis on institutional change and risk situations attempts to verify Adams's argument. As already mentioned in the introduction, with independence Central Asia experienced two major challenges in terms of water management. The first is disintegration of basin management of water, energy and agricultural products, the second is the shift from state and collective farm to farm organisation and water user association. In both cases the actors of the institutions affected by the changes experience situations of uncertainty and risk.

5.1 Risk theories

In the literature risk is distinguished from uncertainty. While uncertainty refers to 'an individual's lack of knowledge about the state of the world', risk is understood when the individual is able to calculate the probability of future events on the basis of experience or scientific observation. (Cashdan 1990, p. 2) Cashdan refers to different variables, to quantify and to measure risk. However for the purpose of analysing institutions in terms of their adaptability to change under the influence of risk it is unnecessary to quantify risk. In this analysis, risk will be defined broadly as hazards and insecurity.

Selin distinguishes between collective and individual risk. However, she refers only to risk in already established reflexive modern circumstances in which the distinction between classes and social structures do not really matter. In these circumstances the economic distinction of risk takers and risk producers does not apply anymore, because the faced risk has a larger scale. However, the distinction could also be analysed in a cultural theoretical sense, if society or institutions are influencing their members on the perception and selection of different risks. (see Douglas and Wildavsky 1982) In such circumstances the risk, which endangers the society or the institution as a whole will be perceived as the greatest risk. Hence, institutions try to focus on managing the risk, which has been socially constructed as the major risk. An institution is more or less stable and unchanging if the prime risk is recognised and accepted. Simply, institutional risk is a situation of insecurity of the institution as a whole. Risk experience is a cultural, social and political construction, hence it can be analysed with cultural, social and political theory.

5.2 Risk and institutional change

How do institutions change in the face of constructed risk? Even though NIE focuses on institutional change it does not take into consideration risk and its influence on the institutions of society or on society as such. However, anthropological studies show how institutions (societies) incorporate new events, events which threaten the existence of these institutions. Douglas using Torry's case study on 'hindu peasant adjustments to famines' shows that institutions at risk do not change their structure or their means of operating. The case study indicates that during uncertainty the institutions protect those in command and those already advantaged. An institutional skeleton is preserved which functions with the usual channels of communication. (Douglas 1986, p. 123) Salhins demonstrates in a case study on Cook's arrival in Hawaii how the Hawaiian society incorporated this event into their normal structure. The old system of cultural-religious rituals and social structure was reproduced. The structure did not change, the new events were interpreted within the old structure. This is similar to Ortner's case study on the Sherpa society. She argues that the society did not change even though the society encountered influences from the outside, which threatened the inner structure. She argues that the structure found solutions to incorporate these events. Even though, the Sherpa had the possibility to change because of the influences from the outside, the 'people could see no alternative to the given order' (Ortner 1989, p. 201) However, the distinction between different levels of institutional structure implies, that not all institutional levels are necessarily at risk at the same time nor do they face the same level of risk or even the same risk at different levels. This could imply that some changes to the structure induced by the awareness of a newly constructed risk get adopted into the system at one level. At other levels they get rejected or manipulated because they endanger the structure of the institution at that level. Sahlins shows in her case study that the way taboos were constructed shifted, but that the main structure stayed static. The structure incorporated new objects, which were to the advantage of the group in power, the ruling elite. The case study in Karakalpakstan shows that even though the old structure is formally abandoned and a new participatory structure is supposed to have taken its place, in terms of power nothing has changed. The old system of power incorporated the institutions and used them to their advantage in terms of land distribution, water allocation and social power.

In modern societies the structure of institutions is based on knowledge and information. In such societies a leading ideology is science, which samples and models reality and attempts to shape the mindsets of the actors to include its ideas and models. Scientists are the elite of the ideology. They interpret new events and risk situations within the framework of science, and are therefore able to evaluate acceptable risks, or not according to whether technical and institutional measures can be mobilised to control or ameliorate the impact of the risk. Because of the shift from pre-modern to modern agricultural production in Uzbekistan during the Soviet era, science and modern technology became first acceptable and now dominating. Because science is dominating the discourse even on the local level, risk situations such as local water scarcity or basin water scarcity are only approached from a scientific perspective, and not from a political or social. The solutions are constrained and political or social alternatives are sanctioned. On a national level Uzbekistan is continuing with agriculture and cotton production even in areas which should have been phased out, such as Karakalpakstan or Karshi.

6 Conclusion

The review of the debate on institutional change in different disciplines has encountered a very diverse set of issues and numerous interpretations. The analysis of actor versus structure in anthropology broadened the perspective of NIE literature. The emphasis of the importance of history, habits and context, with its contradiction between 'inescapable past' and 'irreducible present' as well as the debate on either the individual 'reproducing' a structure or 'choosing' one structure among contradictory options, underlines the question of the sustainability and continuity of institutions and whether institutions are open for change. This study examines attempts to introduce institutional change at a point in history where the polities and the economies in which the reforming agricultural sector was embedded was very destabilised. Integrated policies and administrative entities disintegrated. Hence, the context in which the old institutions were embedded changed. Officially, the institutions are changing as well, at least in name. However, as indicated institutional change is very complex, the analysis will show whether the institutions change and how they change.

The combination of the actor oriented and the structure oriented approach indicates that structures enable and constrain agency. However, every interaction has to be interpreted in the context of time and space. The discourse between actor and structure is significant for the understanding of the problem of collective action and for the creation of common property regimes. One could argue that depending on which approach is taken common property regimes either seem to fail or to succeed. However, choosing an actor oriented approach one has to keep in mind that actors function in a dynamic system. Parameters such as social heterogeneity, leaderships, technology and technological change, markets and outside opportunities interact, to describe them in isolation does not reflect their dynamics and might lead to misinformed policies resulting into a new 'tragedy of the commons'. The creation of water user associations on the local level, is the creation of collective action over a common property resource. Depending on the dynamics between structures and actors these newly created institutions might become sustainable. The case study fills some of the information gap, and provides information about the dynamics between structures and actors on different levels.

The evidence suggests that the need for institutions is based on their function. Institutions not only reduce negotiation and discourse costs but also uncertainty, because they provide a structure for the interpretation of everyday life. However, institutions have also negative effects, because they reflect and reinforce power relations and therefore can be exploited by elites to serve their interest. Both of these functions of institutions are evident in the study area. Because of the old structure the new water user association managers have the power to distribute water. Their scientific knowledge legitimises their position and their decisions. Hence, negotiation costs about water allocation are low. On the other hand the old structure influences the new setting and reinforces old power relations. Therefore the former state and collective farm manager in the role of the new farm organisation manager have authority which they can use to their interest.

The debate on institutional change has shown that not all institutions are equal, and that hierarchical structures exist. The hierarchy is based on the transaction costs of change. The main reason for change has been identified as the lifting of the 'veil of ignorance'; more and new information alters relative prices and therefore induces change. As the case study will show change can be induced from the supply and the demand side. The first changes, such as land reform and the creation of water user associations are supply induced, however the following changes are demand induced, such as training for the members of the water user associations.

The analysis verified that institutions are not static and that they are able to adapt to new situations. According to the findings, institutions and their ability to change is determined by the power of influential stakeholders, which enable or reject changes, hence they function as gate keepers. Their position enables them to ignore or reinterpret new events, which might threaten them or the structures of the institutions. The reinterpretation of new events empowers the stakeholders and the whole institution to deal with the new challenge without endangering the position of the stakeholders and without changing the institution. There were some clear examples of such social process in the study area. For example the disintegration of the state and collective farms and the introduction of farm organisations was reinterpreted from the stakeholders as a continuation of the old structure. The behaviour of the local elite did not change. On the other hand, at the provincial level the disintegration of the basin management framework, changed their perception of the institutions and their positions within the institutions. Their interpretation of the risk situation enabled them to use the setting of the institution to create new stability for themselves.

The findings have significance for policy making and management institutions. Power structures have to be understood and accommodated in the policies of development. Bottom-up movements are bound to fail if they ignore the power structures within society. The management of common pool resources (such as water) can be managed on the local scale only if the main stakeholders are incorporated in the making and implementing of policy. Ideally, these structures have to be mediated so that old stakeholders do not have the chance to take advantage of the new institution to reinforce their power position. Mediation could come from the grassroots but also from top-down initiatives from the government. A strong state is needed to enable these discursive processes. The top of the hierarchy has to be able to reach out to the local level. The local level communities have to be empowered to be critical and question established power structures and to be able to find new alternatives to the current practice.

The theoretical framework of institutional change will be utilised to analyse the shifts from state and collective farm to farm organisation and water user association in Uzbekistan. The framework will help structure events and actions and will help explain the behaviour of actors, such as the local former employees of the state and collective farms, head of hydrological units of the former state and collective farms, state and collective farms and province managers of water distribution. As mentioned in the introduction the two case studies interact and it is not possible to understand the results of the one administrative level without being aware of the influence on other levels. Because of the dynamics between the administrative levels the case study is a very useful analysis with which to verify and to establish a theory of 'institutional change'.

Chapter three: From past to present: agriculture and politics in Uzbekistan

1 Introduction

The purpose of this chapter is to introduce the Republic Karakalpakstan and the Khorezm province in Uzbekistan and the water management constraints that they have faced, and continue to face, because of their downstream location. A second purpose of the chapter is to give an historic overview of society and irrigation in Central Asia from the Russian conquest to independence. The main focus is on the local water management and how it changed through history, and how different institutional settings influenced local management. Attention will be given to the group, which was responsible for water management during this period. It will be shown that the wider context influenced the local setting and the farm.

The chapter is structured into two main sections – first, the description of the region and secondly the history of irrigation in Central Asia, more specifically of Uzbekistan. The historical section is again structured into three subsections addressing the three administrative levels, namely Russia, Soviet Union and Uzbekistan. While the main focus of the first two historic sections is on the local level irrigation and on the elites, which existed at that period, it also provides insights into how irrigation expanded and the influence this had on water management and environment. The last historic section focuses on Uzbekistan after independence, it gives an indication of the current discourses on water use and water policy.

2 The region

2.1 The basin

The Aral Sea Basin covers about 1.8 million km^2 within seven states: Afghanistan, Iran[13] and the five Central Asia Republics, Kazakhstan, Kyrgyzstan, Uzbekistan Tajikistan and Turkmenistan. With the exception of Kazakhstan, the Central Asia Republics lie almost entirely within the Aral Sea Basin. The basin can be described as a large drainage system which terminates in the Aral Sea (Map 3.1). The western and central parts are covered by plains (the Kara Kum and Kyz Kum deserts); the eastern part is occupied by large mountain ranges. The mountain ranges form the flow generation zone for Central Asia's main rivers, the Amu Darya and the Syr Darya, which cross the deserts and flow into the Aral Sea. On their way to the Aral Sea the rivers not only cross international boundaries but are also used as boundaries between the states. Because of the study area the focus will be on the Amu Darya only.

2.2 The Amu Darya

Even though, the upper watershed of the Amu Darya is situated mainly in Tajikistan, the river originates in Afghanistan on the glacier in the Vakjdjir Pass, close to the border of Pakistan's Northern Territories. Up to the confluence with the Pamir river the Amu Darya is called the Vakhan-Darya. From the Pamir to the confluence with the Vakhsh river it is called the Pyandj, and after that the Amu Darya.

---

13 Iran's contribution to the flows in the basin is entirely in streams, which end in the Kara Kum desert, and cannot actually reach the Aral Sea.

Map 3.1 The Aral Sea Basin

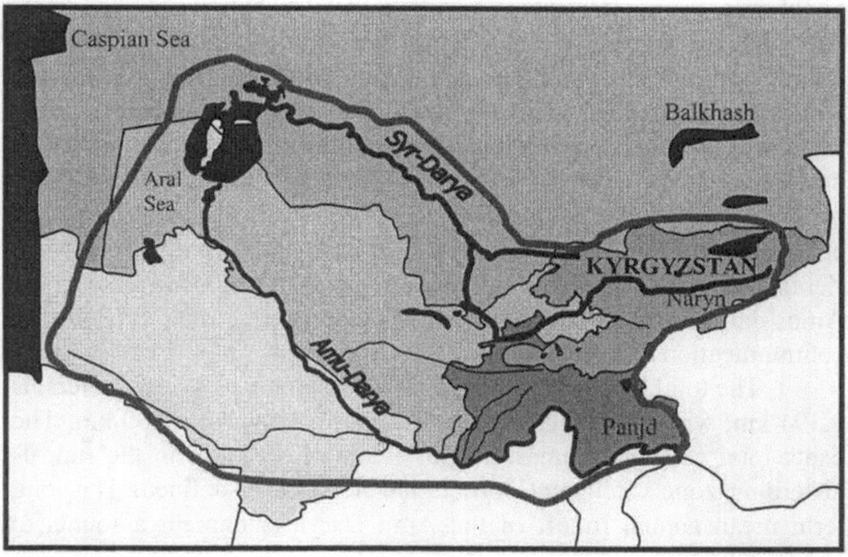

Map 3.2 The Amu Darya

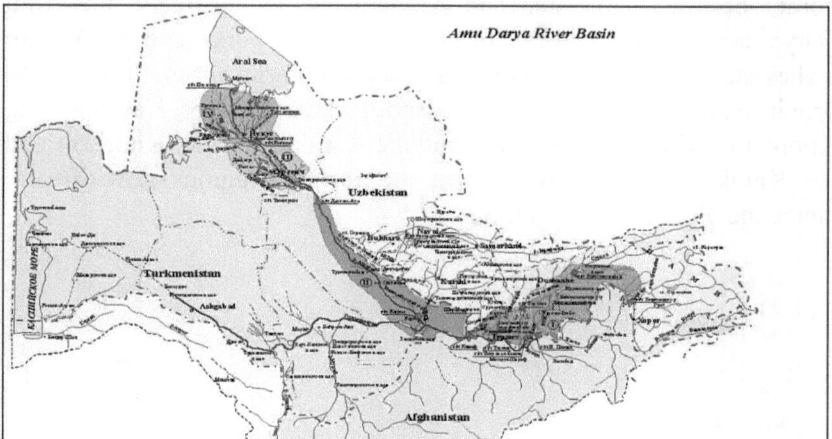

The Pyandj has numerous tributaries, the principal ones being the Shakhdara, Gunt, Bartang, Yazgulem and Vakhch from Tajikistan. After the confluence of the Vakhsh and Pyandj, the Amu Darya is joined by three further tributaries, the Kunduz (from Afghanistan), the Kafirnigan (from Tajikistan) and the Surkhandarya (from Uzbekistan). There are no other major tributaries reaching the Amu Darya further downstream. However, tables provided by SIC ICWC and SANIIRI present data which show that the Zerafshan and Kashkardarya contribute to the Amu Darya as well. Berkoff points out, 'the Zerafshan and Kashkardarya do not strictly speaking contribute to the Amu darya, except through residual drainage water'. (Personal communication D.J.W. Berkoff 27.06.2003)

The total length of the Amu Darya, including its head waters, is 2,394 km, while the length of the Vakhsh river is about 650 km. The exact total catchment areas are not available, because in the runoff-absorbing zone catchment borders are not clearly defined. The long term mean annual runoff of the Amu Darya catchment amounts to about 78 km^3.

The Pyandj and the upper reach of the Amu Darya form the border between Tajikistan and Afghanistan. Downstream, the Amu Darya crosses the territories of Uzbekistan and Turkmenistan. Within Uzbekistan, The Amu Darya and its tributaries pass through the provinces of Surhandarya, Samarkand, Bukhara, Navoi, Kashkadarya, Khorezm and the autonomous republic Karakalpakstan. Khorezm and the Karakalpakstan Republic are the furthest regions downstream, hence the closest to the Aral Sea.

Figure 3.1 The Amu Darya Flow Chart

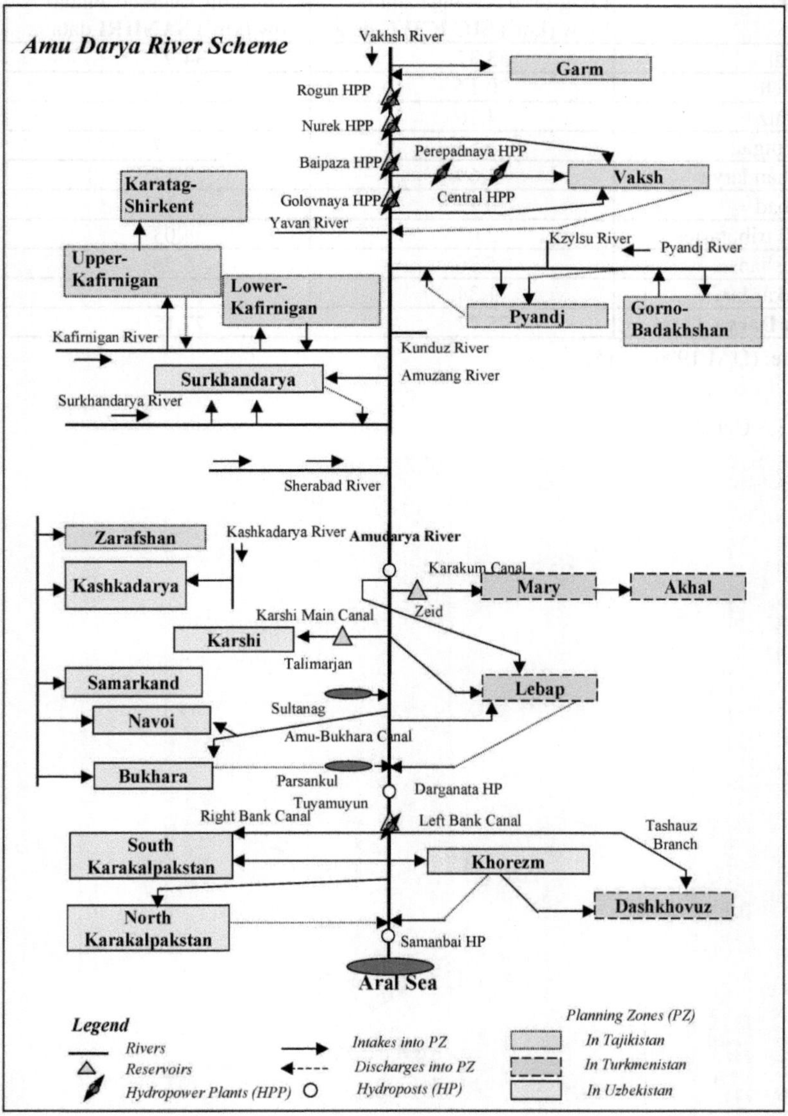

Table 3.1 Average flow according to rivers (km^3)

River	Long-term Average annual Flow (km^3) **SIC ICWC** data	Long-term Average annual Flow (km^3) **SANIIRI** data
Pyandj	33.35	34.92
Vakhch	20.14	20.14
Kunduz	4.16	4.16
Kafirnigan	5.53	5.57
Surkhandarya	3.69	2.17
Sherbad	0.23	0.17
Small tributaries	2.02	4.03
Zerafshan	5.26	5.26
Kashkardarya	1.34	0.73
Amu Darya Total	**75.72**	**77.15**

Source: (DAI 1996, p. 15)

Map 3.3 Uzbekistan

Because the Amu Darya has tributaries originating in different countries and crosses different state boundaries, it is important to know the amount of water originating in one country and the amount of water utilised in that country.

Table 3.2 Average flow contribution[14] according to countries (km^3)

Country	**Amu Dar'ya**	**Syr Dar'ya**	**Total**	**Water used in agriculture**	**Water used in industry**
Afghanistan	6.18 [15]	-	6.18	-	-
Kazakhstan	-	4.5	4.5	27.41	6.26
Kyrgyzstan	1.9	27.4	29.3	9.5	0.59
Tajikistan	62.9	1.1	64.0	10.96	.91
Turkmenistan	2.78	-	2.78	23.29	0.49
Uzbekistan	4.7	4.14	8.84	54.37	3.68
Total	**78.46**	**37.14**	**115.6**	**125.53**	**11.93**

(WB 1996, p. 37)

The table of the World Bank does not incorporate domestic water use, DFID provides the following numbers for Uzbekistan.

14 The problem with the table is that it provides flow data for the natural river basins however the river basins have been altered. Hence, one could argue that the rivers flowing into the Karakum canal, which is part of the Amu Darya, also contribute to the river basin. Hence it would have to incorporate rivers from Afghanistan and Iran flowing into Turkmenistan into the Karakum Canal.

15 According to Ton Lennaerts from USAID NRMP the water flow contribution from Afghan territory to the Amu Darya river is 9 km^3 annually (e-mail discussion from 6 Jun 2002).

Figure 3.2 Sector water use in Uzbekistan (in per cent)

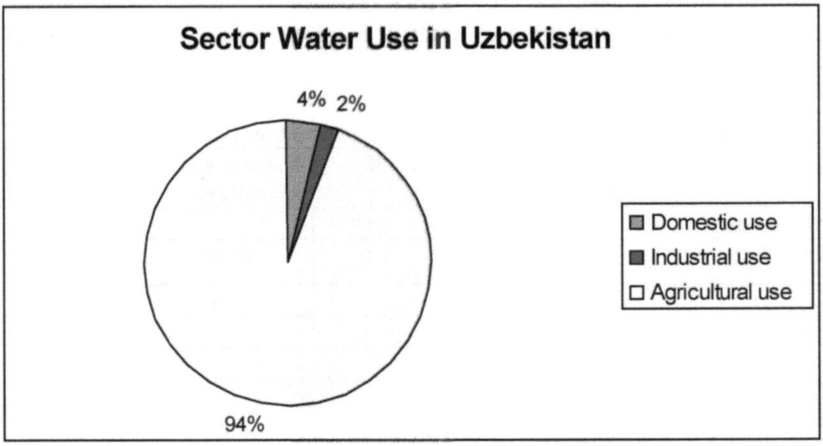

(DFID 2001, p. 17)

The table illustrates and the pie-chart confirms that the agricultural sector is the main user of water. Sokolov explains further,

> the majority of total average withdrawals (93.3 per cent) is consumed by the agricultural sector, with 91.8 percent by irrigated farming. The balance is used for domestic and drinking purposes (3.5 percent), industrial (2.1 percent) and fisheries (1.1 percent). (Sokolov 1999, p. 108)

In Uzbekistan agriculture is mainly based on irrigation.[16] In 1989 4.3 million hectares were irrigated, that is about 82 per cent of cultivated land. The irrigated land produces more than 90 per cent of crop production. Of the total irrigated area, 56 per cent is situated in the Amu Darya river basin. (Compare SIC ICWC 1999, p. 41) The agricultural sector of the downstream regions of Uzbekistan, the Khorezm province and the Republic Karakalpakstan, are dependent on

---

16 The history of irrigated agriculture started more that 6000 years ago in the seven natural oases within the territory of Uzbekistan.

water from the Amu Darya for irrigation. The focus of the case study is these two downstream regions.

2.3 *The Khorezm Province and the Republic Karakalpakstan*

In Uzbekistan the irrigated tracts can be distinguished between old and new areas under irrigation. Sokolov states that 'new irrigated areas have been developed after 1960 with lined channels, tubes and flumes in the on-farm network, and subsurface drainage system. [...] Old irrigation areas still have earth canals [and] open drainage systems'. (Sokolov 1999, p. 109–110) The difference between the two is that the new irrigated areas resulted from the period of the Virgin land expansion. Downstream regions have still earth canals and open drainage systems, if the drainage systems are still functional.

Map 3.4 The study area: South Karakalpakstan and Khorezm province

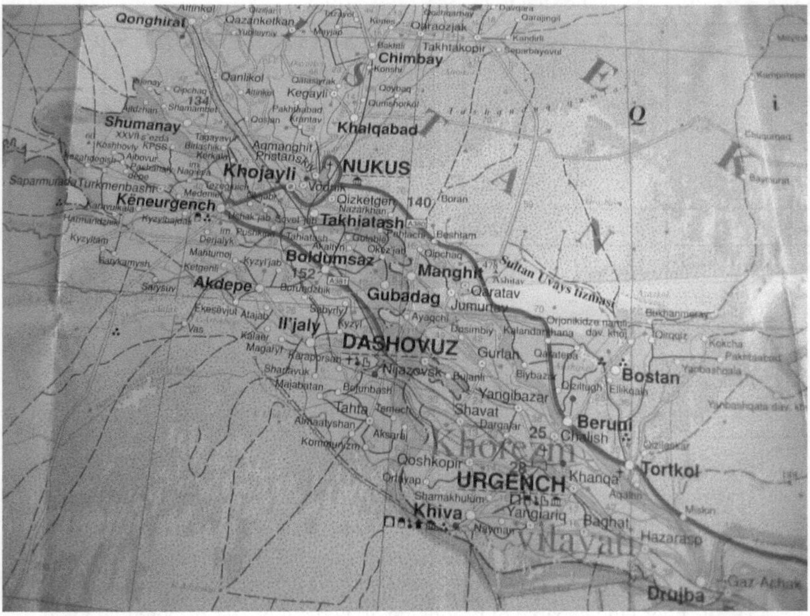

In the era of the Soviet Union, the downstream regions in Uzbekistan and Kazakhstan were specialised in rice production, and were one of the main rice-producing regions in the Soviet Union. (compare Zonn 1999, p. 167, for Karakalpakstan informal interview with Mr. R. Koshekov July 2001) After independence the specialisation in Karakalpakstan did not change, it continued to be Uzbekistan's largest rice producing region. Because of their specialisation, the water allocation to the downstream regions were high. According to Koshekov 12 km^3 was annually allocated to Karakalpakstan. Zonn (1999) points out that in the downstream regions of the Amu Darya the irrigation rate for cotton should be in theory 7,000m^3/ha, however in the Khorezm and Dashkhovuz regions it reached up to 30–40,000 m^3/ha and in some regions of Karakalpakstan it reached up to 50–55,000m^3/ha. Berkoff contests these high figures. (Personal communication D.J.W. Berkoff 27.06.2003) The excessive water use had consequences, such as high salinity in the irrigated areas.

Table 3.3 Fertility of Land Suitable for Irrigation, 1996

Province / Republic	total gross area (000ha)	Total area suitable for irrigation (000ha)	Quality of land suitable for irrigation[17] (% of irrigated land)				Present irrigated area (000ha)
			I 81–100	II 61–80	III 41–60	IV, V < 41	
Khorezm	605	240	40	26	19	15	262
Karakalpakstan	16,659	1,540	0	7	39	54	502

(DAI 1996, p.35)

17 Quality of irrigation is defined as:
I: very good land capable of producing 81–100 percent of potential yield
II: good land capable of producing 61–80 percent of potential yield
III: moderate quality land capablie of producing 41–60 percent of potential yield
IV, V: poor land capable of producing 40 percent or less of potential yield

As the table indicates the irrigated area in Karakalpakstan and Khorezm differ in their fertility and suitability for irrigation. Sokolov states that in the downstream regions especially in Karakalpakstan 'about 95 per cent of the lands are saline' (Sokolov 1999, p. 110). Despite high salinity and reduced yields these areas are still cultivated. In addition to the low quality of land there are problems of water quality and quantity. The metering stations (gauge stations) along the Amu Darya provide data of the salinity and quantity of the flow. The Water Resources Management and Agricultural Production in the Central Asian Republics (WARMAP) project states data for the years 1988 and 1989. The criteria for choosing these two years is that 1988 was a typical year with a high water availability and 1989 a year with low water availability.

Table 3.4 River salinity

Gauge	Distance from Sea (km)	1988		1989	
		Salinity (ppm)	Load (M ton)	Salinity (ppm)	Load (M ton)
Termez	1,277	538		734	
Kerki	1,045	788	37.1	1,098	28.5
Darganata	611	761	29.1	1,163	23.3
Tuyumayun	450	891	31.5	1,108	18.5
Kipchak	287	961	22.4	1,339	12.3
Samanbai	215	923	8.5	1,463	1.6
Kyzldjar	102	958	13.9	1,184	0.8

(WARMAP 1997, p. 56)

The high levels of salinity can be explained through the collector drainage flow from the upstream areas in Turkmenistan and Uzbekistan. The estimation for the drainage flow is 'at 6.5 km^3 with salinity levels that often exceed 5gr/l.' (Hutchens et al 2000, p. 7) According to the World Bank the average discharge of salts is 84 million tons per year. (World Bank 1998) The salinity level of the river during dry years is even higher. During dry years it can exceed 2 gr/l.

In the year 2000 and 2001 Central Asia experienced a drought. The data of the metering stations indicate that the available water declines the further downstream the station is located. The percentage of the water used between the stations is higher then the water allocated to these regions. The data shows that in 2000 and 2001 10 per cent and 1 per cent respectively reached the closest metering station to the Aral Sea. Hence, all downstream regions, but especially Karakalpakstan, suffer from high salinity levels of the agricultural land, saline irrigation water and uncertain amounts of irrigation water annually.

Table 3.5 Water availability downstream during droughts 2000 and 2001

Average Water Discharge							
Gauge	**Distance from Sea (km)**	**2000**			**2001**		
		Actual	**Norm**	**% of Norm**	**Actual**	**Norm**	**% of Norm**
Kerki	1,045	11871.0	21057.0	56.37	10576.0	21057.0	50.23
Darganata	611	7683.0	14170.0	54.22	6158.0	14170.0	43.46
Tuyumayun	450	4416.0	11849.0	37.27	3623.0	11849.0	30.58
Kipchak	287	2731.0	7699.0	35.47	1514.3	7699.0	19.67
Samanbai	215	517.1	3195.5	16.18	34.5	3195.5	1.08
Kyzldjar	102	320.9	3003.6	10.68	32.0	3003.6	1.06

(data available from metering stations)

The presented data gives indication of the broader geographical background and specific conditions on agricultural land and water quality and quantity of the downstream regions. The data on water availability and on the salinity of water and land is especially relevant because it gives some indication of the productivity of the downstream farms and their returns to applied irrigation water. As will be argued in chapter four the reform, which established water user associations is connected to the productivity of the former state and collective farms. Only unproductive farms were privatised and organised as water user associations, and these water user associations were until 2001only in the downstream regions.

However, to understand better the process of institutional change, of privatisation and the establishment of water user associations it is necessary to understand the historical context as well as the geographical and hydraulic context. The next section focuses on the historical, cultural and social perspectives in which irrigation agriculture developed in the region.

3 Historical perspective

3.1 Russia's policy of cotton production

Prior to the Russian conquest of Central Asia in the second half of the 19th century, agricultural production of cotton was already established in the Central Asian region and cotton was traded with the Russian state. In the first half of the 19th century cotton, yarn and fabrics accounted for two-thirds of the goods imported from Central Asia. (compare Pierce 1960, p. 163) Between 1840 and 1850 Russia imported a total of 100,000 poods of cotton from Central Asia, but between 1850 and 1860 these imports rose to a total of 270,000 poods. (Pierce 1960, p. 164) The majority of the cotton imported to Russia had been imported from North America. During the American civil war 1861–65 cotton imports from North America ceased, and Russia had to find ways of getting the necessary cotton for its rapidly developing industry. The decline of cotton imports from North America led to increased imports from Central Asia. Hence, 'in 1861 Russia imported 152,000 poods of Asiatic cotton, and by 1864, 704,000 poods.' (Pierce 1960, p. 164)

Central Asian cotton became the focus of Russian politics. Lipovsky argues that cotton was one of the reasons for conquering Central Asia from 1864–1885. (Lipovsky 1995, p. 529) After the conquest, the Russian government promoted actively the production of cotton in Central Asia.

> Once subsumed into the Russian Empire teams of agricultural and engineering experts were sent to the region to assess its agricultural potential, particularly for large-scale cotton cultivation. It is estimated that between 2.2 and 2.5 million ha of land were being irrigated at the time (O'Hara 2000, p. 369)

Pierce states that, the cotton production in Central Asia at the time of the Russian conquest was still hindered by defects in quality and means of production. (Pierce 1960, p. 164) 'The native Central Asian plant had a very short fibre and was inclined to be very dirty.' (Lipovsky 1995, p. 530) This led to the decision to introduce American cotton and American processing methods in Central Asia. American cottonseeds were of better quality and produced higher yields. Its production took off in the early 1880s and with the introduction of the Trans-Caspian Railroad the rate of exports increased.

Matley argues that the irrigation techniques of the peasant farmers were inadequate.

> Although the Central Asian farmer knew the basic principles of irrigated agriculture, the lack of tools and equipment limited his effectiveness. Levelling fields was difficult, and in uneven ground, hollows received too much water and hillocks too little. Farmers never undertook the drainage of surplus water from the fields, and salinisation and the development of swamps resulted. (Matley 1967, p. 276)

Matley's argument goes against the common thesis, which argues that the irrigation method prior to the Russian conquest and Soviet collectivisation was environmentally friendly and sustainable. O'Hara describes the traditional irrigation system, she states 'in sum traditional irrigation and water distribution systems tended to be small, but had high productivity, were well managed, extremely efficient and sustainable over the long term.' (O'Hara 2000 p. 379) Micklin focussing on a study of irrigation systems at the time of the October Revolution in 1917, reasons that the old systems had been environmentally friendly and sustainable. Micklin argues that in the past, small-scale farms have managed irrigation systems with an average size of 2–3 hectares and with irrigated fields of 0.3–0.8 hectares. (Micklin 1992, p.270) However, Berkoff is suspicious of Miklin's argument, he reasons that 'yields in the newly developed areas are typically higher than in the old lands'. (Personal communication D.J.W. Berkoff 27.06.2003)

The introduction of better seeds was not the only policy which led to the increase cotton production in Central Asia. The government also promoted the production of cotton through economic incentives. It introduced high taxes on imported cotton. Furthermore, 'the government gave tax privileges to cotton growers, taxing cotton no higher than wheat and barley and less than rice'. (Pierce 1960, p. 166) Lipovsky states that this tax privilege was only for land planted with American cotton. However, Uzbek farmers were still hesitant to shift to the new seeds. Pierce argues that the provincial administration from Samarkand 'distributed free seeds of the American variety, and permitted buyers to give cash advances to growers'. (Pierce 1960, p. 165)

The increase of cotton production coincided with the indebtedness of peasants. Even though, the profits for American cotton were higher than food crops the costs of cotton production were higher as well. In addition, with the shift to cash crops peasants were exposed to a wider market with external influences. Indebted peasant farmers lost their lands to a rising class of large landowners. However O'Hara states that 'large cotton plantations along lines similar to the USA were not successful because of the huge labour requirements associated with growing cotton'. (O'Hara 2000 p. 374) As a result she argues, 'plantation owners quickly went bankrupt.' (O'Hara 2000 p.374) According to Pierce the percentage of landless in Ferghana province reached 30 per cent in 1912. (compare Pierce 1960, p. 170) The large number of landless eased the influence of the Soviets over land reforms in the 1920s. (see below)

Overall, the Russian focus on cotton led to a decline of agricultural land planted with food crops and gave rise to a shift from food crops to the cash crop cotton. Consequently, Central Asia ceased to be food self-sufficient. The Minister of Agriculture, Krivoschein, stated in 1912,

> The present development of cotton plantations can and should be intensified still further by means of further reduction in the quantity of grain crops planted on irrigation land. [...] Every extra pood of Turkestan wheat means extra competition for Siberian and Kuban wheat; every pood of Turkestan cotton means competition for American cotton. Therefore it is better to give the territory imported wheat – even at extra cost – but to make irrigated land available for cotton growing. (Lipovsky 1995, p.530)

Hence, wheat was imported to the region at subsidised prices. Central Asia lost its food self-sufficiency and became dependent on outside regions. (Clem 1997, p. 176) By the beginning of World War I more then fifty per cent of the agricultural land used for crop production was used exclusively for cotton. In addition to the shift from food crops to cash crops, did the area of irrigated land expand? Matley argues that the desire 'to increase irrigated acreage was tantamount to a wish to increase cotton production'. (Matley 1967, p. 274)

Throughout the twentieth century until the end of communism the principle of comparative advantage at both the Russian and the international levels was a basis of regional policy and natural resource management. The Moscow perspective intensified the security priorities of the highest administrative entity, the Russian state, and even more so those of the Soviet state in the era which followed. The twentieth century was dominated by the certainties of the hydraulic mission (Swyngedouw 1999, Reisner 1984, Allan 2001) with serious consequences for natural resources. (see below for the hydraulic mission in Central Asia and its environmental consequences)

According to Pierce the Russian expansion of cotton production was not based on an increase of irrigated area, but solely on a shift from food crops to cash crops. Pierce states that the measures taken, 'expanded the cotton growing area in Central Asia, but only at the cost of other crops which were needed for food'. (Pierce 1960, p. 166) This argument is repeated by Lipovsky arguing even though the Russian Ministry of Agriculture declared the

> necessity of bringing new land under irrigation for cotton production. The Russian government, however, withheld its consent to concessions for the construction of irrigation systems from both Russian and foreign businessmen, in the belief that the irrigation of large areas of Turkestan was a matter to be decided at national and state level. (Lipovsky 1995, pp.532–3)

O'Hara contradicts Lipovsky and argues that there were plans for increasing the irrigation area, evidenced in 'M.N. Ermolaev's 1908 plan to divert the Amu Darya westwards to irrigate an additional 565,000 ha of land in the Kara Kum Desert and F.P. Morguchenchov's 1915 plan to irrigate an additional 1.4 million ha of land between the Amu Darya and Caspian sea by diverting the Amu Darya along the course of the old Uzboy River channel' (Zaharchenko, 1990; Sakisov, 1992 in O'Hara 2000, p. 369). Even though O'Hara argues that plans existed to expand irrigation this does not imply that it was anticipated or feasible at that time to put this plans into action.

Even though, Pierce argues that the cotton crops only expanded at the expense of other crops; he shows that the Russian state tried to reclaim former irrigated areas and to expand the irrigated area in Central Asia. Pierce states that in two regions of Central Asia irrigated area were reclaimed: in the Hungry Steppe southwest of Tashkent and in the Transcaspian province. However, the anticipated results did not coincide with the real outcomes. 'Neither of these fulfilled the original hopes of their designers or the great expectations of those who had foreseen vaster achievements encompassing the entire region.' (Pierce 1960, p. 181) It could be reasoned that the poor results from the expansion of the irrigated area, led to a change in policy and was the reason for the statement of the government. (see the quotation of Pierce above) On the other hand, Matley states, 'it is certain that the irrigated area of Central Asia expanded during the Russian period before 1917' (Matley 1967, p. 272) Citing different reports in which the figures of expanded area vary from 815,000 acres to 1,968,000 acres. (Matley 1967, pp. 272–3) In addition to Matley's argument, O'Hara states that during the Russian period the area increased from 2.2–2.5 million ha to 3.2 million ha by 1913. (Pankova, Aydarov, Yamnova, Novokava & Blarovolin, 1996, quoted in O'Hara 2000, p.

370) However O'Hara argues that the overall increase in the amount of land irrigated during the Russian period was modest and generally involved the expansion of existing irrigation systems. Taking all arguments into consideration it seems that the main increase of cotton production was due to a shift from food crops to cash crops and only to a lesser extent to the increase of irrigated area. It is evident that the Russian state started to transform agriculture in Central Asia. However, the outbreak of World War I 'forced a prolonged postponement' and delayed a complete transformation.

3.1.1 Changes of local water management

When the Tsarist troops arrived in Central Asia agricultural production was based on peasant farms with small irrigation plots. Jones Luong states that Central Asian communities differentiated themselves from each other through clans and tribal lineages as well as on the basis of slight cultural and linguistic differences. (Jones Luong 2000, p. 2) Matley (1967), who studied the irrigation management systems prior to and during the Russian conquest describes the local Central Asian irrigation system in the following way. Water allocation and the operation and maintenance of the irrigation systems was decentralised and governed at the village level. Canals and ditches irrigated individual private fields and they were the responsibility of the farmer. The cleaning of the main canals was a communal task. During the winter season families had to supply labour to maintain and improve the communal irrigation systems. Water allocation and distribution to the fields was managed at the village level by a water controller (mirab). The responsibility of the mirab was to ensure 'that each landholder received the flow of water for the number of days to which he was entitled'. (Weinthal 2002, p. 77) In addition the mirab was responsible for the supervision of the construction and maintenance of major canals and for the system of distribution canals and irrigation ditches. According to Matley, the elders from the villages, which used the same main canal, controlled the mirab, so that water was distributed equally, and local upstream – downstream conflicts were reduced. However, he admits that the system was not perfect. He states that bribes ensured supplementary supplies of water and head-enders diverted more than their fair share of water.

Weinthal argues that the Russian strategy was not to dismantle local institutional structures, but to leave them intact not only for administrative purposes but also as a mechanism of social control. Hence, the village elder was still a key person for the regulation of everyday interaction in the agricultural sector. She argues that 'Tsarist administrators retained local irrigation practices to foster settled agriculture and provide for the planting of cotton in the oases.' (Weinthal 2002, p. 75) She continues 'Russia permitted and moreover relied on the former local judicial and political institutions of the village to govern everyday life. Thus the village leaders were able to function in their traditional role as locally elected officials'. (Weinthal 2002, p. 77) O'Neill confirms Weinthal arguing that

> following the Russian conquest of the 1860s, Tsarist rule in Central Asia did not bring an overnight transformation, but was rather the start of a gradual period of change. Some traditional practices, (those relating to water, for example), were hardly touched by the new administration. (O'Neill, forthcoming, p.3–4)

Both arguments confirm Williams' (1967) reasoning. He states that the Russians had a 'Laissez-faire approach to local water management' and argues while 'land tenure, taxation, agriculture and other aspects of the local economy wrought by the Russian arrival' changed, these changes 'were not matched by any kind of organised approach to change water management.' Berkoff gives a possible reason for the Tsarist regime's behaviour, he reasons 'traditionally, these were inundation canals which means that the amount of water diverted is a function of water levels with only very limited control at the intake structure. There were no barrages or pumps and minimal river training at the intake'. (Personal communication D.J.W. Berkoff 27.06.2003) Hence, 'Local custom was left alone, but the context within which the custom had developed was irrevocably altered' (compare Williams, 1967, cited in O'Hara 2000 p. 374–5)

On the other hand Thurman points out that the Tsarist administration 'attempted to introduce reforms in the irrigation sector, but these failed and the authorities declared that irrigation would be run "by custom". However, subtle changes took place "irrigation officials became part of the Tsarist civil service" and, "were no longer hired, fired and paid for" by water users' (compare Thurman, 1999, quoted in O'Hara 2000 p. 374) O'Hara states that the situation was 'exacerbated by the imposition of irrigation officials unaccustomed to the traditional methods'. (O'Hara 2000 p. 374) Hence, the link between water users and water providers was undermined. In addition, according to Thurman the salaries for the irrigation officials were low, and encouraged bribery and corruption in the allocation and distribution of water. O'Hara states that the intensification of competition and the weakening of traditional organisation resulted in an increase in water theft. Hence, the Russian changes in water management were subtle compared to other changes in the agricultural sector. However, the Russians did influence local water management. As mentioned above, the focus on cotton changed the structure of the land. Larger farms developed. It has not been mentioned whether the new farms were owned by local people or by immigrant Russians. If the latter was the case, then the local traditions of water management would have been altered, because it is likely that the newcomers had a different value system and may not have accepted the local practices of mirabs and elders. However, it is also questionable whether the change in land holding size had a similar influence on the local practices. Arguably, the new large farm owners could have opposed the mirabs' or elders' decision on the equal allocation of water. Overall one can assume that the shift of ownership and land size might have also contributed to a weakening of the old irrigation institutions and the rise in water theft and bribes.

3.2 *The period of the Soviet Union prior to World War II*

3.2.1 *Land reform and dekulakisation*

During and directly after World War I Central Asia suffered from food crop dependency on Russia. Starting in 1914 food supplies were cut partially, and in 1918 they stopped completely. Due to the resulting shortages, peasant farmers shifted back from cash crop to food crop production. In addition to the change of production, the irrigated area also declined. The decline was caused by anti-communist partisans, who seized food supplies and destroyed the irrigation system. Matley states that 'by 1918 the irrigated portion of the Turkestan ASSR, fitting approximately within the boundaries of the old Turkistan guberniia, was only 37.5 per cent of the total area irrigated in 1915.' (Matley 1967, p. 285) However in the period from 1918 to 1920 the irrigated area increased again 'to 3.3 million acres or 50 per cent of the 1915 figures'. (Matley 1967, p. 285)

During the 1920s (from 1922 to 1928) one emphasis of the Soviet authorities was to restore the irrigation systems. Already in 1918 Lenin signed a decree, which allocated 50 million roubles for irrigation development works in Turkestan. (Zonn 1999, p. 158) In 1928, 82 per cent of the irrigated area compared to 1914 had been restored. The new Soviet administration focused on the production of cotton. In 1927 the area devoted to cotton was slightly larger than in 1914. (Matley 1967, p. 287) Micklin argues that in the mid-1920s 'so-called "modern" irrigation techniques were mandated in Central Asia by the communist regime.' (Micklin 1992, p. 270) However, it is questionable whether this mandate could have been implemented immediately, especially in a time of radical change in land ownership.

The Bolshevik Revolution and the beginning of the Soviet Union brought a period of radical change. In December of 1925, the communist party of Uzbekistan declared that all land and water was state property, and set limits on holdings, above which land would be confiscated and redistributed to the landless. Stalin introduced a programme of collectivisation and cotton independence, however as

Thurman points out, 'economic reality mattered little to the Stalinist leadership, whose priority lay as much in simply controlling the villages and effecting social transformation as in reform agricultural production.' (Thurman, 1999, p. 9) Carlisle argues that in 'Uzbekistan the first Five Year Plan brought an assault on traditional classes and created new social clusters. The early victims were the Muslim version of Russia's kulaks' (Carlisle 1991, p. 99) Thurman confirms this, also mentioning collectivisation as the main task. 'Collectivisation was also combined with other policies of social transformation, such as campaigns to wipe out Islam and traditional family roles. (Thurman, 1999, p. 13–14) On the other hand O'Neill argued, 'the Bolsheviks intended to eradicate "class". [...] This took precedence over any notion of agricultural advancement or equitable land tenure.' (O'Neill forthcoming, p. 17) Therefore, one could argue that not collectivisation and economic efficiency was the priority of the plan but the political and economic control and social transformation in the villages.

The 'dekulakisation' undermined existing structures of authority in villages and served as a convenient pretext for eliminating anyone opposed to Soviet policy of collectivisation. Hence, dekulakisation shook up village power structures. However, as Thurman points out it 'was an arbitrary process that could be used by activists or villagers for decidedly non-class motives such as acquisition of property and/or setting personal scores.' (Thurman, 1999, p. 16)

3.2.2 Collectivisation

Weinthal argues that the early attempts of collectivisation failed in Uzbekistan. She states that in 1924 there were only 62 kolkhozes in all of Uzbekistan. (Weinthal 2002, p. 83) This is confirmed by Thurman who states that the farming population opposed collectivisation. Weinthal reasons that after the first failure of collectivisation the 'Soviet authorities initiated land and water reforms between 1925 and 1929, making newly reclaimed land available to those willing to accept collectivisation.' (Weinthal 2002, p. 83) According to her the number of kolkhozes increased by 1927 to 832. It seems that the process of collectivisation was not refused after the reform, however, Thurman states that those

> refusing to 'volunteer' to join the kolkhoz were threatened with an impressive array of reprisals. These included arrest, use of armed force, exile as a bay, loss of the right to vote in local meetings and elections, and denial of land, water for irrigation, industrial goods and/or food. (Thurman, 1999, p. 14–15)

The importance of water delivery as method of compliance is pointed out by O'Neill, who argued that water was the

> 'main weapon' used in the break up of the traditional tenure system and the subsequent seizure of lands [...] Using the threat to withhold water supplies as a sanction, they were able to force compliance with what was becoming a 'command' agricultural policy. (O'Neill forthcoming, p. 17)

Hence, the increase in kolkhozes was forced, and not voluntary. In addition O'Neill argues that the period of dekulakisation and redistribution to sharecroppers weakened the agricultural sector. He reasons that the new land owners had too small plots and only a few were able to survive from their plots. Hence, the dekulakisation prepared the ground for the subsequent revolution of collectivisation. O'Neill argues

> In the heat of dispossessing their landlords, many sharecroppers had sealed their own fate by smashing machinery, destroying seed, and killing or driving away the landlords, thereby depriving themselves of the equipment, capital and expertise they would soon need as owner farmers. (O'Neill forthcoming, p. 18)

The result was widespread famine, and a 'population in no position to resist when their newly bestowed lands were collectivised a few years later.' (O'Neill forthcoming, p. 19) O'Neill's reasoning indicates that the new peasant farms did not have any other alternative other than complying with the collectivisation, and if they did the compliance procedures of the new administrators were too strong. This might explain why in Uzbekistan the collectivisation was ahead of schedule. 'By the end of 1932, 77.5 per cent of the rural households in Uzbekistan had been incorporated in 9,734 kolkhozy and 94 sovkhozy.' (Thurman, 1999, p. 14) However, resistance took place, passive resistance was more common than confrontation.

A form of passive resistance was such that farmers often slaughtered their livestock rather than surrender them to the state. One of the consequences was widespread shortage of draft power, fertiliser, meat and milk. This could be one explanation why the economic success of the collectivisation was limited. In 1928 the collective farms produced less than peasant farms. 'Yields per hectare on collective farms were 25 percent lower than on independent farms.' (Thurman, 1999, p. 11) In addition passive resistance continued when farmers became kolkhoz members. The resistance was expressed in low labour productivity, pilfering of farm property and produce and reporting false production figures.

By 1928, 1,360,700 hectares were under irrigation in Uzbekistan. 'Of the irrigated area, 44.7 per cent was under grain, 38.9 per cent under cotton, and 9.2 per cent under alfalfa.' (Matley 1967, p. 288)

> During the first Five Year Plan, work began on extending irrigation of such areas as the Farghana Valley, and the Mirzachol Sahra, and land along the Zaralfhan, Surkhan Darya, Chichiq, and Ahangaran rivers. In all, about 200,000 hectares of new land was irrigated between 1929 and 1932. (Matley 1967, p. 294)

By 1932 the irrigated area had increased to 1,776,500 hectares. 'The area under cotton had increased to 55.9 per cent of the irrigated land, while the area under grain had plunged to only 15.8 per cent.' (Matley 1967, p. 288) About 61 per cent of all cotton fibre produced in the Soviet Union was raised in the Uzbek republic.

The agents of collectivisation were mainly activists from the European part of the USSR, however indigenous rural activists, encouraged by Moscow, were also involved. Thurman states 'since the

KpUz was still weak in rural areas of Central Asia, the bolsheviks relied on activists like the '25,000ers' Communists' from Europe. (Thurman, 1999, p. 15) These alien activists and local plenipotentiaries who implemented collectivisation subsequently became chairmen or directors of the collective farms.

3.2.3 *The new elites*

The management of the collective farms was influenced through the strong-arm approach to creating them. According to Thurman 'The Stalin regime's penchant for using terror to impose its will upon the countryside would carry over into the management of kolkhozy after collectivisation.' (Thurman, 1999, p. 15) Here a second reason for the low performance of the collective farms emerged.

The goals of controlling rural areas and fomenting social revolution came first, and organising farms second. (Thurman, 1999, p. 15) However, the new social system was not too different from the old structure. One the one hand the collective farms were formed from a single kinship group or clan. (Jones Luong 2000, p. 10) Jones Luong confirms the observation of Winner who argued that new superstructures were grafted 'onto pre-existing villages or onto groups with similar kinship structures.' (compare Winner 1963 quoted in Weinthal 2002, p. 84)

The position of the village elder and of the mirab was replaced by soviet farm institutions.

> Rural state and Party nachal'niki (managers or bosses) replicated in a new guise the time-honoured elites of an older authoritarian order. Soviet 'water lords' who controlled the irrigation system, represented a modern version of an ancient profession and practice. (Carlisle 1991, p. 99)

In addition Jones Luong points out, that 'collectivisation disempowered traditional leaders (i.e. tribal chiefs, or beys in Kazakhstan) by giving "control" over the use and distribution of land to kolkhoz (collective farm) chairmen.' (Jones Luong 2000, p. 10)

The policy undermined the authority of the traditional leaders, because it eliminated their source of authority. In addition because the tribal structures and identities of pre-Soviet Central Asia lacked any official recognition or institutional resources 'as a result, they were effectively disempowered and the strength of their ties was diluted over time.' (Jones Luong 2000, p. 10) Jones Luong goes even so far to reason that 'the collective farm chairman effectively usurped the traditional role of the clan leader'. (Jones Luong 2000, p. 11) The result of dekulakisation and collectivisation was that a soviet-oriented rural aristocracy was imposed on the traditional order. Even though, the soviet planners wanted to abandon the pre-existing forms of patronage relations, in reality they reinforced them.

One main difference was, that prior to soviet rule these elders were elected in the village, with the new system, the candidates were appointed from above. 'By law, kolkhoz members could elect their own boards and chairmen, and govern the internal affairs by majority vote in meetings. However, the Party controlled the process of appointment and "candidates" in elections were merely the favourites of the district executive committee.' (Thurman, 1999, p. 23)

The district authority controlled also the funds and production of the collective farms. 'Expenditure of internal capital funds, acquisition of inputs, and "sale" of kolkhoz produce were similarly dominated by district authorities.' (Thurman, 1999, p. 23) In addition, the machine-tractor stations were also on the district level and furthered the dependency of the local level on the district. Jones Luong contradicts Thurman, according to her it was not the district but the regional level, which was the source of power. She reasons that

> Through his position as the chief executive in a given territory, the obkom first secretary became the primary dispencer of political and economic resources at the regional and local level and skilfully used this position to build loyalty and support throughout his oblast. (Jones Luong 2000, p. 9)

Jones Loung again mixes her reasoning of political and economic resources with indigenous Central Asian tradition. She reasons

> this vast political and economic authority vested in the obkom first secretary had the unique effect of supplanting the authority of tribal and local leaders. The institutionalisation of this authority alone formed a "natural" basis for the re-definition and extension of existing clan and tribal based patronage networks to the regional level.' (Jones Luong 2000, p. 10)

It seems that the system of patronage was reinforced at different levels, at the district and at the provincial level as well. Hence, a superstructure of patronage was created. Weinthal argues similarly. She reasons that the policies of collectivisation and centralisation furthered the system of patronage, and also extended the system to higher levels. This new superstructure of patronage was

> associated with the kolkhoz (collective farm) and the sovkhoz (state farm) at the micro level and with administrative-territorial districts at the macro level associated with the rayon [district] and the oblast [province]. (Weinthal 2002, p. 84)

Because of the links between micro and macro levels, water management and agriculture was controlled from above. O'Hara argues 'benefits were no longer linked to duty and water users had no say in how the system was managed nor were they responsible for its maintenance.' (O'Hara 2000 p. 379)

The members of the state and collective farms at the operational level did not have any legal right to decision-making in agriculture and water management, consequently they lost interest in agricultural production and irrigation efficiency. O'Hara sums up the situation after collectivisation, she states the Soviet built systems were 'huge, inefficient, inflexible, poorly managed and for the most part, unsustainable.' (O'Hara 2000 p. 379)

3.2.4 *Tackling inefficiency?*

The social and institutional problems caused a decline of the agricultural production in the 1930s period. The collectivisation of peasant farms coincided with a further increase of the irrigated area, by 1934 933,000 hectares. The new areas were dedicated to cotton production. However, collectivisation resulted in a sharp decline of cotton production. Although the irrigated portion of Uzbekistan increased steadily during the early part of the collectivisation drive, by 1933 and 1934 signs of strain began to appear. In 1933 the area planted to cotton had reached a total of 2.4 million acres (968,000 hectares), compared with 1.06 million acres (425,000 hectares) in 1913 on the territory of present day Uzbekistan, the total harvest amounted to 817,000 metric tons. However, though in 1934 the cotton fields amounted to 2.3 million acres (933,000) the harvest dropped to 738,000 metric tons in this period of very low yields per hectare. (Matley 1967, p. 289)

According to Soviet planners, the dilemma of falling yields, which was caused by social and institutional factors, had technical solutions. The Soviet planners started 'a campaign for increasing productivity per acre by applying more chemical fertiliser and planting a better quality of cotton seed'. (Matley 1967, p. 290) While in 1932 65,000 tons of chemical fertilisers were shipped to Uzbekistan in 1937 it reached 569,000 tons. In addition seed selection centres were established in Ferghana and in Khiva.

In addition to the technical solutions, it seems that during the mid-1930s the social policies of the Soviet system also changed. Carlisle analysed the Soviet internal politics in terms of action and retreat. He reasons 'by the mid-1930s, the dekhans (peasants) continued to practice the old ways while meeting the state's economic requirements, for which the most productive were amply rewarded by high cotton prices.' (Carlisle 1991, p. 99) According to him

> the powers-that-be had learned that the essential economic goals of increased cotton production and state crop procurement could be achieved without

trampling on or directly attacking traditional Muslim society. (Carlisle 1991, p. 99)

It seems that the focus on reforming agricultural structure, has shifted from a prioritisation of social reform to a prioritisation of agricultural production. However, the change in focus allowed that the old structures of patronage could re-emerge. The re-emergence might even have increased the patronage relations on the local and above local levels.

Starting from the 1930s the agricultural policy of the Soviet Union focused on bringing virgin land under agricultural production. (Pallot 1995, p. 104) However, not until 1939 did the first large-scale irrigation projects since the pre-1917 period begin. (Matley 1967, p. 294) In the beginning of 1939 45 canals were built, the longest was the Great Farghana Canal (168 miles) and in 1940 the North (100 miles) and the South Farghana (65 miles) canals followed. (Matley 1967, p. 294–5) Matley states that 'although a few canals were completed during World War II, the next extensive irrigation projects did not take shape until after the war. The irrigated area in the Mirzachol Sahra expanded between 1939 to 1941 from 95,000 hectares to 138,000 hectares.

3.4 *The Soviet Union after World War II*

3.4.1 *The Soviet 'hydraulic mission'*

The Soviet Union was an ideal administrative structure to implement what has come to be called the hydraulic mission. Those public institutions that launched hydraulic missions in the capitalist neo-liberal West and the socialist Soviet Union were inspired by the same assumptions and the same tendency to focus state power and public allocative capacity to developing natural resources to enhance the state's economy and the social conditions of rural communities. The unifying assumption was that natural resources and particularly surface water resources could be controlled. Civil engineering, hydraulic technology and the accumulation of investment resources could all be combined to reshape the environment to create storage and flood protection works that would enhance the economy and rural conditions. Politicians, water professionals and existing and new farmers were all certain that this was the way to approach water resources. In the West the green movement began to challenge the certainties in the 1960s (Reisner 1984) and persuaded a US president by the mid-1970s that certainty should be replaced by uncertainty. (Carter 1982) In the case of Central Asia the move to uncertainty and reflexive policies considerate of the water services provided by surface water did not take place. As a consequence the Aral Sea Basin environmental outcome became the tragedy that has been widely reviewed.

During the hydraulic mission the irrigated area was not only expanded, the land was also transformed for mechanised farming. In 1950 a new Resolution 'On the transition to the new system of irrigation for the more comprehensive use of irrigated lands and improvement of agricultural operations' was adopted. 'The Resolution called for a major reduction in the number of irrigated plots and an increase of their area up to 20, 40, 60 and 100 ha.' (Zonn 1999, p. 160) By 1950 the cotton area had passed the 2.5 million acre (1 million hectare) mark. The policy of bringing new areas under irrigation reached its height with Khrushchev's project of 'virgin land' in 1953. The project was supposed to raise agricultural productivity. By 1956 an additional 88.6 million hectares of land was cultivated in the Soviet Union, mainly in Kazakhstan and Western Siberia. As part of the 'virgin land' project Khrushchev promoted the idea of expanding the irrigated areas in Central Asia. (Rumer, 1989, p. 88–9) In 1956, under the 'virgin land' policy and the beginning of the 'hydraulic mission' the area under irrigation expanded further and reached in 1960 180,000 hectares, with the aim to have reclaimed 600,000 to 700,000 hectares by 1965. (Matley 1967, p. 295–6)

With the increase in irrigated area the area allocated to cotton increased further. By 1963 the area allocated to cotton was 4.1 million acres (1,628,000 hectares) with a total production of 3,689,000 metric tons. (Matley 1967, p. 291) In the Brezhnev period (1964–1982) irrigated land was further expanded. In this period the II and III stages of the Karakum Canal as well as the Karshi and the Amu-Bukhara canals were commissioned. The headwater intake of these canals range from 200 to more than 500 cubic m/s. Large reservoirs such as the Andhizhan, Charvak, Chardarin, Tuyamuyun and Nurek reservoir were put into operation. (compare Zonn 1999, p. 166) These structures give indications of the dimensions of the water management works during the period of the hydraulic mission.

Before the construction of the Kara Kum Canal the irrigated area in Turkmenistan was only 0.35 million hectares. The irrigated area increased from 0.5 million hectares in 1965 to a total of 1.25 million hectares in 1988. (Lerman & Brooks 1998, p.165) Kotlyakov states that the irrigated area increased in the early 1960s 'in Uzbekistan and Tajikistan by 1.5 times, in Kazakhstan by 1.7 times and in Turkmenistan by 2.4 times'. (Kotlyakov, 1992, p. 285; also Glazovsky, p. 105) While in 1965 an area of 4.5 million hectares was irrigated in Central Asia, by independence the total irrigated area had increased by an additional 2.5 million hectares. (Kotlyakov, 1991, p. 5) Zonn states on the plans of constructing the Karakum Canal,

> attention was paid only to the engineering aspects of the construction of the canals, reservoirs, etc. [...] no one took into account the damage that would be caused by seepage from the unlined canal, by salinisation of hundreds of thousands of hectares of lands and by the resettlement of villages on new, less fertile land. (Zonn 1999, p. 161)

The increase in irrigated area also led to new settlements of non-indigenous populations. Even though, one could assume that the new settlement did not comply with the institutionalised system of patronage. According to Weinthal the patronage relations continued even on the new settlements. (compare Weinthal 2002, p. 86)

3.4.2 Integrating the management of upstream & downstream flows

Because of central control and basin management the utilisation of the rivers did not correspond to the administrative boundaries and the interests of the administrative zones. During the Soviet era, the Central Asian rivers (Amu Darya and Syr Darya and their tributaries) as well as their regulating structures were managed with an awareness of the river basin. The borders between the republics were only administrational and therefore did not hinder basin management of the Amu Darya and Syr Darya.

The Central Asian water authority was headed by the Ministry of Land Reclamation and Water Resources of the USSR (MLR&WR). All republican institutions and republican interests in resource utilisation were subordinated to the Central Authority in Moscow and to the interests of the greater Soviet Union.

> The ministries of the Central Asian republics were extensions of the ministry in Moscow. They were responsible to fulfil the centralised plans and norms. Their role in decision-making was limited to provide data to the centre. (Renger 1998, p. 5)

The SIC ICWC report states that that the subordination was two-fold, sectoral [irrigated agriculture] and national. (SIC ICWC 1999, p. 13)

Lange explains the sectoral subordination, he states 'the water management infrastructure was designed for a unified purpose and placed where it made sense geologically.' (Lange 2001, p. 1) Within the basin framework the dams and reservoirs were built upstream in the mountains, while the irrigation areas were downstream in the valleys and in the steppes. The water management constructions were built to enhance irrigation in the downstream regions. To use the dams for agricultural purposes, water had to be released in the summer and fall to supply irrigation. The basin framework approach had the benefit of total control over water and efficient water management for irrigation.

In the Soviet system there was no conflict between upstream and downstream interests. Upstream and downstream riparian units benefited from the regional approach and the use of water, energy and food as common pool resources. Due to the forced focus on irrigation the upstream water management structures, such as dams and reservoirs, did not produce hydroelectric power when it was needed the most in the upstream regions, which is during the winter season. The dams released water during the summer when the downstream riparian administrative units needed water for agriculture. Because all the republics were unified as if a single country, energy was provided during the winter from Russia and the downstream regions, which are rich in oil and gas. The soviet system was special and to an extent sustainable because it integrated the management of a number of sectors – food supply, cash crop production (cotton) as well as energy. (compare Wegerich 2002a)

3.4.3 The Aral Sea crisis

A literature review shows that the reasons for environmental destruction in the Aral Sea Basin are manifold. McKinney points out that there are two distinct groups of scholars, identifying two different causes for the environmental destruction of the Aral Sea region. Former Soviet water planning specialists believe that the root cause is an incorrect economic development strategy. (McKinney, 1997, p. 4) On the other hand Micklin, argues that the introduction of modern large scale irrigation techniques and the shift from individual farming towards collective farm enterprises is the cause of the environmental crises. The following analysis indicates that both approaches are valid in that both explain the crisis to some extent.

One of the main representatives of the first approach is Kotlyakov. He argues that 'the region's economy has reached a catastrophe, not because of the desiccation of the Aral; conversely, the Aral catastrophe is the inevitable result of a deep crisis in the regional economy.' (Kotlyakov, 1992, p. 287) The Soviet economic strategy for Central Asia focused on a high specialisation in agriculture, especially in cotton production. Cotton is a high water consuming plant.[18] However, the specialisation in cotton production was not the main reason for the increased depletion of the Aral Sea. The collected data shows that since the 1960s the area of irrigated land has increased rapidly. Berkoff points out, that 'the primary reason for the reduction in the water reaching the Aral Sea was the massive expansion of the irrigated area onto the intervening steppes. As a result, not only was evapotranspiration greatly increased but much of the drainage flows failed to return to the rivers and ended up in shallow desert lakes. The Karakum canal takes 50 per cent of the Amu Darya at the diversion point and is perhaps the main cause'. (Personal communication D.J.W. Berkoff 27.06.2003) A third reason for the high demand of water is the use of low-productivity land, which required more water. (Kotlyakov, 1992, p. 286) Glazovski agreeing with Kotlyakov, adds that the development of cotton production focused on the expansion of irrigation systems and not on increasing the productivity of already irrigated land. (Glazovsky, 1995, p. 106) However, the increase in the expenditure on fertilizers and pesticides which was on a genuinely higher level in the Central Asian Republics compared with other Soviet Republics could lead to the conclusion that an increase in productivity on already irrigated land was achieved.

Arguably, the policy focused on an increase of the productivity of the land and not the increase in water efficiency. This is validated by Kotlyakov who argued that individual irrigation systems tended to be of low quality design, construction and operation. (Kotlyakov, 1992, p. 286) Glazovsky substantiates Kotlyakov's reasoning with estimates showing that 30–45 percent of water was lost due to canals and drainage collectors being constructed without linings. Glazovsky stresses that irrigation techniques were 'primitive', including the use

18 Berkoff argues, 'cotton is not a particularly high water user although it is in the ground for longer than wheat or other crops. It uses little more water than an alternative wheat/pulse rotation'. (Personal communication D.J.W. Berkoff 27.06.2003)

of furrows or even watering by hand instead of the use of sprinklers. (Glazovsky, 1995, p. 107) On the other hand Berkoff argues, 'sprinklers for cotton are out of the question in Central Asia and Glazovsky is misleading. Even in far smaller US surface systems, furrows irrigation was (and often still is) the preferred option. Glazovsky may be correct that 'old' canals were not lined, but many were (e.g. on the Karshi and Hunger Steppe) and these systems were well laid out and complemented by full drainage systems. The issue was not really whether they were inefficient in water terms but whether they were economically justified'. (Personal communication D.J.W. Berkoff 27.06.2003)

Although Micklin (1992) admits that large-scale expansion of irrigation is responsible for the drying up of the Aral Sea, he argues additionally that the problem is deeper and that concentrating on the current irrigation system is a superficial way of addressing the real issue. For Micklin the introduction of large-scale enterprises and the collectivisation of private land are the fundamental causes for the destruction of the Aral Sea.

The literature shows that the Aral crisis was the result of a combination of environmental factors, inappropriate water allocation and management policies and institutional deficiencies.

3.4.4 Soviet elite recruitment: from ideology to skills

In his analysis of political elites in the Soviet system Welsh argues that the traditional system of elite recruitment shifted from an ideological to a managerial and technocratically based system. Hence, 'education has replaced ideological orthodoxy'. (Welsh 1979, p. 124) Putnam confirms the rising importance of technical knowledge in elite recruitment in the Soviet system. In the field of land reform and water management, Putnam states, 'the education of the newer recruits is concentrated in technical fields, particularly agronomy and engineering.' (Putnam 1976, p. 209) The reason for the shift in recruitment is the enhancement of the mobility of the elite, and the reduction of the dilemma of the problems of adverse selection and moral hazard. Welsh's argument of shift in the recruitment of technocratic elites contradicts Winiecki's view that 'the appointments are made primarily on the basis of loyalty rather than managerial competence'. (Winiecki 1996, p. 67)

According to Welsh the leadership in the soviet system changed. 'The changes in political leadership in socialist systems may be traced to one of two considerations: first, the inevitable pressures towards pragmatism, flexibility, and regard for technical and managerial expertise created by economic and social modernisation; or second, the conscious grouping of political leadership groups for the more effective and efficient means of governance, i.e., the desire to retain cohesive, hierarchical power structures in the face of increasing pressures of a disintegrative and pluralistic character'. (Welsh 1979, p. 123) However, this does not mean that the patron-client relationships in the party apparatus disappeared. Putnam reasons, 'the path to the top now requires technical skills and managerial experience, but the path itself still lies within the party apparatus'. (Putnam 1976, p. 210)

The changes in elite recruitment have changed the mobility of the elites in the institutions. Putnam argues, 'the specialised elites display growing self-consciousness, growing aspiration for autonomy and influence on policy making, and growing divergence from the policy views of the party officials'. (Putnam 1976, p. 210) Welsh distinguishes in his analysis the difference between specialist elites in general and elites in the party, he argues 'Soviet specialist elites are substantially more instrumental and substantially more participatory than are their counterparts in the party apparat'. (Welsh 1979, p. 138) Furthermore, he distinguishes between local and central party apparatchiks, arguing, 'local party apparatchiki are more participatory and instrumental than their counterparts in the central party apparatus. That is the local party people are more similar to the specialist elites in their orientation'. (Welsh 1979, p. 138)

3.4.5 Forming a new elite

Prior to World War II coercion to meet the cotton plan came mainly from the secret police. 'Dehqans were jailed, exiled, and even executed for failure to meet the cotton plan, criticising the party agricultural policy or simply getting on the bad side of a well-connected kolkhoz chairman.' (Thurman, 1999, p. 38) Thurman argues that after World War II growing party presence in the kolkhozy replaced terror as a means of coercion. He reasons that by 1952, there were primary party organisations in all kolkhozy in the republic and that rural soviets were given an expanded role in agriculture after Stalin's death in 1953. According to Thurman 'training programs were set up. Waves of specialists who also happened to be party members were sent to occupy leading positions on farms.' (Thurman, 1999, p. 38–9) However, it seems likely that training programs were connected to party membership, hence if individuals wanted to receive education, they had to be party members.

It is pointed out by Thurman that 'between 1953 and 1959, the number of specialists on collective and state farms in Uzbekistan grew by a factor of more than four.' (Thurman, 1999, p. 39) However, one could also argue that the number of specialists on the farms was more connected to the idea of the time, which was a modern worldview which believed in the control of nature by science and state organisation. Hence, on the basis of this view it would have been necessary to train farmers to receive the highest results in irrigation, cotton production and farm management.

Thurman argues that 'By the late 1960s, party penetration of the kolkhozy extended down to the level of field brigades.' He continues 'By 1970, 8.1 per cent of the labour capable population of kolkhozy in Uzbekistan were party members, including almost all farm chairman and directors, as well as large share of the managerial and technical staff.' From this observation Thurman draws the conclusion that 'The party had fused its structure tightly with that of agricultural production in order to maintain control in the villages.' (Thurman, 1999, p. 39) However, one could also reason that the system succeeded, in institutionalising agricultural production, hence the system formalised an acceptable structure. This structure was based on specialists with special knowledge about different issues in agricultural production. Hence, the system managed to transform agricultural production from a pre-modern to a modern framework. This modern framework needed specialists, which became the new rising elite. Through this framework the patronage system could continue to exist within a modern setting.

Critchlow's analysis differs from Thurman's, instead of evaluating the influence of the party in agricultural production he examined ethnic factors.. He observes that while in the beginning of the collectivisation mainly European activists were in influential positions in the agricultural sector. Over the years the percentage of officials in high positions of Central Asian origin increased. He states that

> in the Uzbek Central Committee Secretariat during the 1940s, Russians and other European party secretaries and department heads constituted three-fourths of the total, and no change in this proportion was apparent in the 1950s. (Critchlow 1991, p. 27)

'By 1966 the number of Russians and Europeans in these posts was not three-fourths but approximately one-fourth of the total and remained fairly constant in ensuring years.' (Critchlow 1991, p. 27) To be fair, the indigenous population had a demographic advantage. More young Central Asians finished as graduates of higher education. These graduates started to be the core of scientific cadres. Critchlow states that the number of scientific workers with Muslim nationalities increased between 1960 and 1969 from 17,000 to 48,000. (compare Critchlow 1991, p. 28) Critchow refers to the new scientific workers as Muslim elites. He argues that 'The new elites could be said to owe their status to the Soviet regime, which had given them modern professional training and an economic and social structure in which to use it.' (Critchlow 1991, p. 29) He interprets the influence not as much as a rise in party power and membership, but in a shift from pre-modern to modern system. However, the modern system of professionism met a cultural system 'where tribal memories were recent, where the rank-and-file looked instinctively for leadership to a chieftain and his council of elders'. (Critchlow 1991, p. 19) This discrepancy led to a system in which the reign of the new elites had a patrimonial character.

3.4.6 Elites and hierarchy

Critchlow argues that the new professional elite 'suffered from a paradox of Soviet society: the system encouraged minority education but restricted the occupational mobility of the Uzbek and other Muslim graduates towards the Centre.' (Critchlow 1991, p. 29) According to Jones Luong this is due to the 'scarcity of trusted cadre'. A consequence of this lack of trust was that 'comparatively few Central Asian elites were ever promoted to serve in Moscow, or anywhere outside their titular republic' (Jones Luong 2000, p.11) Restricted to their republic, local leaders primarily vied for promotion to positions at the regional-level in their own oblast, while the highest office regional leaders could realistically hope to attain was in the republican-level government. Jones Luong argues that 'this had an unintended effect on the development and reinforcement of regional political identities.' (Jones Luong 2000, p. 11) Locked into their own region they could build and maintain a viable local power base to advance their career.

In addition the economic specialization had the consequence that many specialists from the same region were trained in the same institutes and then returned to their regions. According to Jones Luong 'this created yet another layer of camaraderie upon which to build regional bonds.' (Jones Luong 2000, p. 11) The specialisation by region and province contributed to the practice of long tenure for elites within the same region or the transfer between similar regions. Again this fostered regionalism and power networks.

While Jones Luong sees the regional level as the main concentration of power and argues that the republican level is more weak than one would expect. This conclusion is based on a study of tax contribution and tax evasion from different provinces. (presentation at Columbia University 2002) In addition she argues that

> during the Soviet period all agriculture was placed under the control of regional leaders. In this context, the republic-level officials acted essentially as a 'mediator' or 'broker' between Moscow and the various regional leaders to extract the maximum amount of production. (Jones Luong 2000, p. 10)

Jones Luong argues that the provincial levels had the most influence and power, she states that 'As in the rest of the former Soviet Union, over time, the creation of regional administrative units institutionalised a pivotal leader – the oblast party committee (obkom) first secretary – with the discretion to select and replace local party and government personnel in both political and economic spheres as well as to '[check their] performance ... [unify] and [co-ordinate]' their activities and '[procure] ... supplies' to ensure their optimal performance.' (Jones Luong 2000, p. 10)

On the other hand, Weinthal does not even distinguish between republican and regional elites. Critchlow, agues from a top down perspective. He reasons that

> the long tenure of the Central Asian first secretaries enabled them to put their personal stamp on the republican machinery as in a fiefdom, appointing their followers to senior posts at republican, oblast and rayon levels. (Critchlow 1991, p. 19–20)

The regional levels were dependent on the republican level, however this lower levels leaders became 'entrenched in their sub-fiefdoms, creating their own entourages.' (Critchlow 1991, p. 20)

According to Jones Luong the different province leaders were in competition with each other. This was already inherent in the structure. At the republic-level parliaments and Communist Party congresses oblast first secretaries and other regional leaders were expected to articulate and represent the interests of what was considered their constituency. Jones Luong states that

> The primary function of republic-level parliaments, for example, was to determine budget priorities and expenditures for each oblast within the republic, which forced regional leaders both to compete and compromise over the allocation of resources.
> Similarly, at party congresses, regional leaders characteristically 'attempted to increase the resources available for the development of their [particular oblasts] by altering the priorities assigned to issues on the policy agenda or even by altering the agenda itself. (Jones Luong 2000, p. 13)

The result was 'not only [to] pit regional cadres [from the same republic] against the established priorities of the center, but against each other, as well.' As resources became more scarce in the 1980s, 'the evidence suggests [...] that competition among regional elites, as well as between regional representative and the center, [...] increased.' (Jones Luong 2000, p. 13) Jones Luong's argument is in direct contradiction to Weinthal, who argues that the republican leadership controlled many of the main posts and/or political and economic appointments within the republics.

The implications of the above processes were that the republican leadership could penetrate and affect societal relations and establish firm social control over a large segment of society. (Weinthal 1998, p. 161) According to this reasoning it was the republican leader who had control and who could have used their influence to change appointment on the provincial level, in case the provincial leader would act against the republican level. On the other hand Weinthal argues in a later publication that

> Concerning the first level, Moscow needed cheap cotton extraction. The leaders of the various republics supplied this through the regional leaders in exchange for a certain degree of freedom over internal affairs. Here the regional leaders in particular benefited. As long as they 'brought in' good cotton harvests, they could control strong patronage networks through the resources they managed. (Weinthal 2002, p. 97)

However, one could interpret the different levels not in a hierarchy but in positions of influence and different bargaining power. But this could be true for different levels and would not only be a restriction to republican and provincial level. Hence, it might be possible to argue that Jones Luong's and Weinthal's latest reasoning could be used on different levels, such as the local and the district level as well. At the district level the state or collective farm managers bargained for resources, district allocations as well as for shifts in priorities for their level as well.

Jones Luong argues that the provincial leaders had special positions because they were entrusted to select candidates in local and republican elections. According to her the basis of this trust was that they were supposed to 'the most knowledgeable about the quality of local cadre'. (Jones Luong 2000, p. 13)

However, between local and provincial level was an intermediate level, the district. It seems unlikely that the district authorities did not have any power in terms of choosing candidates for lower levels themselves. Even though Jones Luong argues that the provincial leaders

> enjoyed the bulk of influence on the electoral process through their direct supervision over the activities of the District Electoral Commission (DEC), in which the Soviet system concentrated authority over the conduction and results of the elections. (Jones Luong 2000, p. 13)

It does not mean that the district level had not the possibility to manipulate the outcome or to influence the choices of the provincial level. One has to keep in mind that the provincial level acted through the district level, and not directly through the local level. Hence, each administrative level had influence over sublevels and each level had its own system of patronage. Weinthal looking at cotton production argues 'In order to ensure that production goals were met, Moscow, however, needed to rely on the republican leaders, who in turn needed to depend on the oblast [provincial] leaders, who then had to entrust the rayon [district] leaders to procure the cotton from the heads of farms. (Weinthal 2002, p. 99) In her statement Weinthal shows the interdependence of the different administrative levels, each of them relied on the lower level and had to trust that the top-down orders were followed.

The 'Uzbek affair' or 'cotton affair' of the 1980s shows that the patronage system affected all levels. Critchlow states

> while societal transformation at all levels was the ultimate design, the elites had become the immediate target, for until they could be made to collaborate effectively in bridging the policy-implementation gap between directives of the central authority and the masses, far reaching change was doubtful. (Critchlow 1991, p. 41)

However, at the time Moscow saw the local corruption as the most serious. Hence, the anti-corruption campaign 'sought to break the hold of the networks of local officials who over the years had built a power base that, despite nominal subservience to central authority, enabled them to stave off change from outside.' (Critchlow 1991, p.41) Here, one would have to question whether the elite at all levels was able to starve off top-down induced changes. Moscow criticised that 'the "nomenclatura" system, which was still in force, had been perverted by "nepotism", "local favoritism", "kinship", "personal devotion", "obsequiousness" and "protectionism"' (Critchlow 1991, p. 41) The consequence of the cotton affair was that at all levels and not only the local level elites were accused of corruption.

3.4.7 Elites and the patronage system

During the 'cotton affair' cotton production was linked to the system of patronage. But how did the system of patronage function and who benefited? Weinthal argues that cotton production was the foundation of a system of strong social control. She distinguishes between three main interest groups, which were linked through various reciprocal relationships. She reasons,

> the first relationship binds Moscow to the republican leadership, including the regional elites. The second relationship integrally connects the republican and regional elites with the general population. The third relationship ties Moscow directly to the population. (Weinthal 2002, p. 97)

She argues that these reciprocal relationships were based on the extraction of cotton rents. The rents extracted from the sale of cotton supported a system of patronage that, in turn, served as a basis for a system of social control. 'Political elites and even farm heads were awarded with promotions and bonuses for bringing in the cotton harvest.' (Weinthal 2002, p. 99) Hence, the cotton rents reached down to local indigenous clients such as farm chairmen.

But according to Weinthal also on the level below the farm manager the population benefited. She argues that 'even when most agriculture was geared towards cotton, families still had private plots where they could produce their own food along with additional crops for sale in the markets.' (Weinthal 2002, p. 100) This was, she states 'to ensure political and social order and to garner labourers for the cotton harvest by allowing the farmers to concurrently work in the informal economy.' (Weinthal 2002, p. 100) In addition, Weinthal states that 'farmers received subsidised farm inputs such as free water and cheap fertilisers and fuel'. (Weinthal 2002, p. 100) Ilkhamov points out that the state farms provide the households under special rates with power, gas and forage and even allowed them to use state farmland after harvesting for private use. (Ilkhamov 1998, p. 552) Based on these 'subsidised goods' households could produce goods for the local bazaars. The income from these informal activities became the main income of the households. Weinthal states that the republican and regional elites ignored these violations of state goods 'in exchange for bringing in the cotton harvest and moreover, for not protesting about women and children having pesticides sprayed all over them when they were sent out to the fields to glean the cotton.' (Weinthal 2002, p. 100)

According to her reasoning, the patronage system allowed Moscow to implement the agricultural policy of cotton monoculture.

> Although this obviously was a situation in which the hegemon (Moscow) was both coercive and exploitative, it also supplied a public good in the form of stability via an incredibly elaborate patronage system. (Weinthal 2002, p. 96)

Weinthal's third identified relationship is between Moscow and the population. According to her, 'Moscow provided social protection and in exchange, the population did not engage in overt protest.'

However, it is questionable, whether this last relationship between Moscow and the population existed. It is doubtful that the population in the Soviet Union could protest before, and if they did to what extent the protest was suppressed or whether a wider audience was aware of the protest. Arguably this changed with Gorbachev, and the policy of openness and reform of the 1980s. The glasnost and perestroika era gave rise to environmental movements. According to Weinthal the all Union Minvodkhoz (Ministry of Land Reclamation and Water Resources) and the Republican Minvodkhoz provided social protection. She argues that the

> Uzbek Minvodkhoz used these [soviet] subsidies to not only expand the irrigation networks but also to build factories, parks, and entire towns. [...] Minvodkhoz was an immense bureaucracy that in addition to construction projects was responsible for the social services of its employees living in these towns; it operated the hospitals, the schools and the main recreational facilities (Weinthal 2002, p. 98)

Even though Moscow was coercive and exploitative, Weinthal argues that 'the region was characterised by levels of political and social order much higher than found in the Slavic parts of the Soviet Union.' (Weinthal 2002, p. 96) She reasons that this was due to the patronage system of social control through cotton rents. However, as it was argued earlier in the chapter, resistance in Central Asia did not have the form of open conflict, but of quiet resistance. This form of resistance was evident in the case of the 'cotton affair'. Critchlow states that 'at the heart of the cotton affair were charges that Uzbek officials at all levels of the republic had been defrauding the central government through an elaborate system of bribe-taking and padding reports.' (Critchlow 1991, p. 41) The corruption and bribe taking seems to be in contradiction to Weinthal's assumption of the relationship between elites and Moscow. It also puts into question whether the provincial elites had strong control over the districts and local levels (Jones Luong's argument) or whether they had to comply as well to the lower levels to keep their position and maintain the patronage system.

According to Critchlow, the bribes and report manipulations were not necessarily based on dishonesty of individuals and the wish to seek profit. He states that 'Some republican and local officials broke the rules not so much for personal gain as in the interest of their communities.' (Critchlow 1991, p. 45) Hence, the elite corruption had positive sides, it was a form of community protection and resistance. 'It was a factor in boosting the morale of the Uzbeks and other Central Asians, creating at the same time a kind of de facto autonomy that enhanced the role of the local population in handling its own affairs'. (Critchlow 1991, p. 45) According to Critchlow, the corruption and the resistance against Moscow enhanced the legitimacy of the elites. He states

> there is evidence that they were regarded with favour by the masses because of their role in devising extra-legal solutions to pressing economic and social problems, and for forming a native buffer between the masses and an alien authority' (Critchlow 1991, p. 51).

3.5 Uzbekistan since independence

3.5.1 Dis-integration: upstream versus downstream interests

With independence and the shift from a single administrative unit to an autonomous state, the regional approach of water management was at risk. The international 'security and conflict community' assumed that Central Asia would have water disputes and water wars. Smith's statement reflects these thoughts, 'nowhere in the world is the potential for conflict over the use of natural resources as strong as in Central Asia'. (Smith 1995, p. 351) While shortly after independence analysis has focused on potential disputes over water allocation amongst the riparian states, in recent years attention has focused on the conflicting uses of water upstream and downstream and the interdependence between energy and water. Neither the allocation nor the provision problem have been solved. In addition to the problems mentioned is the difficulty of establishing legitimate basin management authorities. (for more detail see Wegerich 2002a)

3.5.2 Cotton and society

According to Jones Luong 'Soviet policies and institutions in Central Asia created, transformed and institutionalised regional political identities.' (Jones Luong 2000, p. 1) Hence, Soviet rule created the potential for a national identity. Jones Luong argues that after independence the regional political identities persisted, this independence 'ensured that the very same actors, interests and the basis for evaluating power asymmetries would continue to dominate decision-making in the post Soviet period' (Jones Luong 2000, p. 1) She reasons that the 'common past continued to influence their subsequent development, and hence, the mechanism for institutional continuity.' (Jones Luong 2000, p. 2) Jones Luong's argument is confirmed by Spoor arguing that the vested interests of the political economy of the cotton sector has affected decision making after independence (compare Spoor 1998, p. 411) Weinthal argues similarly,

> Uzbekistan sought to keep the general population on the farms and engaged in cotton production to ensure their hold on social control and stability. [...] Uzbekistan could not jeopardise the foreign revenue earned by cotton sales abroad. (Weinthal 2001, p. 26)

According to her, the state had to continue with the full control over cotton production and sales because it allowed the elite to reinforce regionally-based patronage networks. (compare Weinthal 2001, p.29) On the other hand Spoor, argues that the state had to continue with cotton production because of the 'social pressures (in particular related to rural employment)' (Spoor 1998, p. 415) and Weinthal states similarly 'radical restructuring of the CA political economy would have been socially destabilising'. (Weinthal 1998, p. 192)

3.5.3 Changing networks[19]

While during the Soviet period the party had influence on collective farm manager positions, after independence, the system changed and informal networks were utilised to receive these positions.

The 22 former shirkat managers which have been asked in the whole of Khorezm answered that it is the hakim who makes the decision on the position. Only one interviewee referred explicitly to the hakim on the province level, and not to the district level, all the other 21 interviewees mentioned the hakim on the district level.[20] From the 22 former shirkat managers asked in the whole of Khorezm province, 20 stated that money is involved in the decision making who becomes a shirkat manager. 18 stated that networks are important to get the position as a shirkat manager. According to the 18 interviewees, networks were based on family and friendship relationships. It was argued by the 18 interviewees who mentioned networks that the price to become a shirkat manager varied, according to the relationship. Family and friends pay less.

---

19 See Wegerich (2003) The data on network structure in Khorezm is based on research conducted in February to May 2003. The data was presented at the Conference 'Governing Environmental Flows' in Wageningen 13 & 14. 06. 2003.

20 According to the law it is the employees of the shirkat, who decide on the position of the manager.

Table 3.6: Network ties of shirkat managers in an upstream district in Khorezm[21]

Khorezm Province	**X1** 96– 99;	**X2** 99–now,	
upstream district	**Y1** 97–2002 studied & worked together with X1	**Y2** (AB)2002–now; friend & (brother-in-law?) & classmates of X2, & related by marriage to province 1 deputy Hakim	
I	**AA**: 98–2002 close friend to X1studied & worked together	**AL**: 2002–now uncle of X2	
II	**AB**: 97–2000 engineer with X2	**AM**: 2001–now; related to X2, was agronom in chief under X2	
III	**AC**: 97–now, friend of a friend of X2, friends by work in cotton plant		
IV	**AD**: to 2000	**AN**: placed by Y1, relative	
V	**AE**: 96–99, placed by X1	**AO**: 99–01, person of X2, neighbour & relative	**AR**: relative of Y1
VI	**AF**: 93–now; placed by 1 Province hakim, relative of city hakim of Urgench		
VII	**AG**: 97–now; friend of X1, relative to Y1 (through marriage)		
VIII	**AH**: 98–2001 relative through marriage with X2,	**AP**: friend of Y1, worked together in Oblast level water department	
IX	**AI**: 98–99 friend of X2, but could not perform	**AQ**: 99–2002 friend of X2, accountant of same cotton plant	**AS**:
X	**AJ**: 93–now, small farm, specialist on cattle, no relative ties		
XI	**AK**: 97–now, relative of X2(through marriage)		

21 Shirkat farms in the upstream district are represented with numbers I to XI, shirkat managers are represented with letters from AA to AS, province hakims with X1 and X2, district hakims with Y1 and Y2.

Money, which all the former shirkat managers mentioned, does not appear in the network table. However, the table showes family and friendship ties of the current shirkat managers and the shirkat managers which were managers during the drought period. The tables of the network ties show that nearly all shirkat managers in this particular district in Khorezm have family and friendship ties with the province or district hakims. The data of the district in Khorezm validates that the party lost its influence when it comes to positions and informal networks have taken over.

3.5.4 Food security

After independence Uzbekistan changed policy on food security. It aimed to achieve food and nutrition security as a development objective. (Babu and Tashmatov, 2000 p. 1) Uzbekistan targeting food self-sufficiency chose different strategies to attain this goal. Firstly, the government allocated more land to rural households. The entitled area increased to 0.25 hectares of irrigated land or 0.5 hectares of unirrigated land per household. The total area owned as household plots increased to 650,000 hectares.

The second step to achieve food security was based on a reform of the cropping structures. According to this policy irrigated area allocated to cotton, was reallocated to grain and other food crops such as vegetables, fodder crops, orchards and wine yards. (Tashmatov 2000, p. 111) FAO data does not validate an increase in area harvested for vegetables or rice. However, the data confirms a decrease in allocated area to cotton and an increase in area harvested for grain. The area allocated to cotton decreased from 1,666,700 ha to 1,425,000 ha in the time period 1992 to 2000, area allocated to grain increased from 626,990 ha to 1,108,000 ha in the same period. (FAO data from 28.01.2002) Tashmatov argues that the increase in grain production was sufficient to achieve grain security, 'which allowed Uzbekistan to stop importing grain and to use locally produced grain'. (Tashmatov 2000, p. 111) Khusanov even states that 'the implementation of the Program of Grain Independence has allowed Uzbekistan to turn in a short period of time from a grain importer into a net grain exporter.' (R. Khusanov 2000, p. 123)

3.5.5 Cotton versus rice

The main focus of the agricultural production was still cotton in the post-independence period. This emphasis was rigorously implemented even during the drought years 2000 and 2001. In this period the downstream regions, received less water compared to the regions which focus on cotton production. The 1999–2001 'drought' had a devastating effect on Karakalpakstan's rice production. Even though in the year 2000 115,000 hectares were planted, 'only 45,000 hectares were harvested'. (ORYZA 2000, November 29, Alimatov, 2000) According to FAO data in the whole of Uzbekistan the rice area harvested in 1999 was 164,000 ha. The area shrank in 2000 to 124,900 and in 2001 to 37,557 hectares, and the total rice production fell from 420,800 in 1999 to 154,800 tons in 2000 and 135,542 in 2001. (FAO data from 28.01.2002) However, it seems that the data of the FAO has its problems, because in October 2001 the FAO data for Uzbekistan showed for 2000 as area harvested 65,000 ha and production of 175,000 tons.

The effects of the 'drought' for the farmers were devastating. In October 2001 an UN assessment mission to Uzbekistan determined that 45,000 people might experience food supply problems. (The Times of CA, 2001) OCHA gave a higher estimate, stating that the water shortage 'has rendered 100,000 farmers (79,000 people in Karakalpakstan, 21,000 in Khorezm) without a stable source of income.' (OCHA No. 2, 2001) Northern Karakalpakstan is the most affected area, and also the furthest downstream in the Amu Darya Basin. According to Romanova, the region's nine northern districts have produced no rice in two years. (Romanova 2001) As a result people have left the farms for better opportunities in other regions. OCHA states that the drought has 'led to limited population movements from the most affected areas to other parts of Uzbekistan, as well as across international borders to Kazakhstan and the Russian federation.' (OCHA No. 2, 2001) Romanova claimed that, 'entire villages began leaving northern Karakalpakstan'. (Romanova 2001)

However, even in the year 2001 the areas in Karakalpakstan planting cotton received water. 'The region's nine northern districts have produced no rice in two years and cotton farming is down by more than a third.' (Romanova 2001) Water was allocated to farms specialised in cotton production. R. Koshekov (head of water use department, Ministry of Agriculture and Water economy of Karakalpakstan) stated that water gets distributed to the districts, which specialise in cotton production. (Informal interview, Nukus, July 2001) Within the districts, the head of water distribution in the district allocated water to the cotton producing farms. (Informal interview with J. Sulaimenov, deputy head of the Department of Agriculture and Water economy Kungrad district, July 2001) The politics of water distribution was not against Karakalpakstan but against rice production. Berkoff argues, 'it is inevitable in a drought that the rice area would have to decline. Less water must be allocated in full to a smaller area rather than spreading the water (as is possible for cotton) since such a strategy would have led to no production at all'. (Personal communication D.J.W. Berkoff 27.06.2003)

FAO data confirmed that the Uzbek government favoured cotton production against rice production. While rice production decreased from 414,000 to 135,542 tons in the period from 1999–2001, cotton production increased from 3,600,000 to 3,790,000 tons in the same period. Furthermore, the cotton production rose even though the area harvested declined. The increased output was due to rising yields.

The central influence on the regional and local departments on water distribution in 2000 and 2001 made sure that farms specialising in cotton received enough water. The upstream regions specialising in cotton production were also not affected by the drought. T. Lennaerts, a specialist from the Central Asian Natural Resource Management Project (NRMP), interpreting the data from the reservoirs argued that the water shortages caused by the 'drought' have been moved downstream. (Informal interview, Tashkent, August 2001)

Cotton remains the main cash earner for Uzbekistan. According to UzbekWorld.com Uzbekistan's cotton fibre exports were worth $897.1 million last year. (UzbekWorld.com 2001) In its report UNESCO states that cotton exports accounts for 40 per cent of the total exports. (UNESCO report 2000, p. 64) The policy during the drought to allocate water to cotton crops instead of all crops indicates that cotton is more valuable for the Uzbek state and its command-administrative system than food crops.

3.5.6 Irrigation management

As illustrated throughout this chapter, agricultural production with its focus on cotton was driven by an ever expanding irrigation system. Currently in Uzbekistan 4.3 million hectares are irrigated. Even though there has been a shift in the dominant water management paradigm in the West from modern to reflexively modern, this shift has still not taken place in Uzbekistan. Hence, the strategy of Uzbekistan is to continue with cotton production or even to expand it. The following three examples will demonstrate the prevailing modern approach, characteristic of modernity, rather than late modernity, to water management, which determines the setting of irrigation agriculture in Uzbekistan.

One of the leading water managers in Central Asia Vadim Sokolov from SIC ICWC argues, that in Uzbekistan

> approximately 634,400 additional hectares are suitable for new irrigation developments, and water conservation would allow a limited expansion of irrigated area, total irrigation potential can be estimated as 4.9 million hectares. (Sokolov 1999, p. 109)

During the time of the Soviet Union a plan was developed to augment the water resources of the Aral Sea Basin by diversion of water from Siberian rivers, via a huge system of canals and pumps stations.

However, because of the environmental implication and the costs of the project the plan has not been implemented. However, Laennerts points out 'that the Representative of the Russian Federation at the Jubilee Meeting of ICWC in Almaty on February 21, 2002 confirmed that the Russian Federation and the Central Asian Republics have renewed talks on the possibility of diverting a small part of the Ob River's flow into the Aral Sea Basin.' (e-mail discussion May 2002) The renewed Uzbek demand for water from outside the basin was further confirmed at a Workshop on Afghanistan in the US in June 2002.

According to a news report from December 2001, the Republic of Uzbekistan has applied for loans and credits in the amount of $140 million towards the cost of rehabilitating and improving the Karshi Pumping Cascade and irrigation infrastructure on the Karshi Steppe. The Karshi Pumping Cascade lifts 200 m^3/s of water up to 160m, essentially for irrigation of 392,000ha in the Karshi Steppe. Even though, the project is not economically viable. However, it is arguable that the current system could break down in the near future and that there is no alternative to agriculture for the inhabitants of the Karshi steppe. However, the same argument could be made for Karakalpakstan during the drought and there the lack of alternatives did not matter.

3.5.7 Aral Sea disaster: where are we now?

The claim that 10 km^3 of water reaches the Aral Sea annually (UNESCO report 2000) is very optimistic. A flow of 23 km^3 is enough to sustain the sea at the current level. To refill the lake to the level of the 1960s the time before the expansion of large-scale irrigation was implemented would require an annual flow of 60km^3. To refill the lake 1000km^3 are needed which would be equivalent to 10 years of full flow of the Amu Dar'ya and Syr Dar'ya together.

The level of the Aral Sea is not the only changed factor. Since the 1960s a variety of other factors changed the possibilities of the current policies and put constraints on the future development of the region.

Table 3.7 Changes from 1960 to 1990

	1960	1990
Irrigated area [million ha]	4.5	7.0
Population [million]	14	50
Water requirements [km^3]	60	120
Salinity of the lake [g/l]	10	40

The present status of the Aral Sea can be defined as unsustainable. The development strategy of the Soviet Union led directly to a condition of environmental unsustainability. The current situation could be defined as beyond sustainability or even as a position in the sphere of an ecological catastrophe. However, the definitions are fuzzy and it is a question of ideology how the situation is defined.

Past developments determine the future development options of the region. It is not possible to turn back the wheel and to focus on the past as a possible future. The governments of the downstream countries [Kazakhstan, Turkmenistan and Uzbekistan] already indicated that they are not inclined to return to the situation of the 1960s. Evidently it is not politically and socially feasible to reduce irrigation to save water for the environment or to cut the production of cash crops such as cotton in a time of economic crisis and foreign currency dependency.

3.5.8 Environmental and population constraints in Central Asia

An exact estimate of the population growth rate within Central Asia is not available. There are only rough estimates. UNESCO claims that the rate of annual increase is currently around 1.61 per cent; the World Bank estimate is by 2,5 percent. Taking the UNESCO rate, in 2025 the population in the region will have increased to 70 million.

As the analysis of the region and the historical data shows under the influence of Russia and later the Soviet Union, Central Asia was forced to focus on an agricultural development based on cash crop mono-culture at any price, irrespective of economic, social and environmental costs. The two countries with the highest environmental burden caused by the production of cash crops through irrigation were Uzbekistan and Turkmenistan. The World Bank reports that Uzbekistan has an irrigated area of 4,280,600 ha, of which 2,140,550 ha are highly saline, Turkmenistan's irrigated area is 1,744,100 ha of which 652,290 ha have high salinity.

In addition to the high level of salinity, in Central Asia the use of fertilizer was very high compared to other Soviet Union republics. In 1965 Uzbekistan used 147 kilograms of fertilizer for each irrigated hectare (kg/ha), this level increased to 238 (kg/ha) in 1975 and 306 (kg/ha) in 1987, compared to an average of 122 (kg/ha) in the Soviet Union in 1987. The amount of pesticides used for agriculture was approximately 30–35 kg/ha, this is almost 30 times higher than the average amount in the USSR. The extensive use caused degradation of the arable land as well as the pollution of groundwater. Both have long term implications for the future development and cause currently health threats to the rural and urban population.

The sudden independence of the Central Asian states left the economic infrastructure in turmoil. Central Asia depended completely on imports in kind and money transfers from the Soviet Union. With independence the Central Asian states had to search for new trading partners for their natural resources and had to make up for the income loss from Moscow.

Even though, Turkmenistan and Uzbekistan are poor in terms of domestic water resources, they are rich on other natural resources. Both countries have large natural gas and oil resources. It is anticipated to utilise these resources for export. However, both countries are landlocked; therefore the costs to reach the world market will be high. Furthermore, neighbouring countries are also trying to export gas and oil; this might lead to competition and to less profitability of natural resource exploitation.

4 Conclusion

Institutions are based in the real world, and are influenced by the conditions of the political and social environment. The first part of the chapter provided the background to the physical world influencing agriculture in the down stream regions. The information on land and water quality and water availability provided a review of the environmental endowment on which agriculture was based. However, as argued in the theoretical chapter the veil of ignorance, protects the institutions, because it covers the facts or finds convenient explanations for the perceived reality. The downstream water scarcity was perceived by the farmers downstream as 'natural' drought. Only officials at higher administrative levels were aware that there were political and institutional reasons for the water scarcity. Hence, different institutional levels had different perceptions of reality and therefore a different perception of risk and uncertainty.

In the light of the argument that history influences the present, and therefore has an influence on if and how institutions change, the history of agriculture and local water management has been evaluated. The historical analysis indicated that prior to the Russian conquest and the later imposition of Soviet Union ideology water was managed at the local level in a system of patronage. As argued, this system has continued throughout the Russian empire and the era of the Soviet Union. Looking at the different statements and arguments a common feature is the patronage system. It was a continuous feature because of the importance of local identity and kin groups, of the formal institutions such as communist party and the related education system, as well as the perceptions of outsider versus insider. Hence, the patronage system in the agricultural sector and in water management has been the basis of the water management institutions in Uzbekistan. Understanding the patronage system is an essential preliminary to understanding water policy and allocative practices in the region. The patronage system will especially influence changes at the local level. This phenomenon will be observed and discussed in the next chapter.

Chapter four: The local administrative level

1 Introduction

In this chapter the top-down institutional changes at the local administrative level, relating to such issues as land reform and the introduction of water user associations, will be analysed with respect to factors that constrain institutional change to which attention was drawn in the chapter which reviewed the theory of institutions and institutional reform. This chapter is divided into two main parts. The first will analyse the first land reform and It will be shown that both reforms varied in their emphasis and scope.

The focus of the study is on local stakeholders and how they complied with the top-down ordered land reforms which started to change the structure of the state and collective farms of the Soviet Union period. The stakeholders at the local level are the employees of the state and collective farms, but also power stakeholders on the local level such as state and collective farm managers and heads of hydraulic units. In addition the influence of stakeholders at higher administrative levels and of the new independent farmers will be considered.

The first part critically reviews the current literature on the first land reform, and describes what has changed and how the processes of change have been reflected in the literature. It is argued in the literature that stakeholders manipulated the reforms. Here this perception is critically evaluated in examining how stakeholders acted during the reforms and what were their individual reasons for their actions. In addition while the current literature focuses on one level of reform implementation, in the first part the linkages between different power levels are introduced.

The second part of the chapter focuses on the second land reforms, which incorporated water reforms. In these reforms farm organisations and water user associations were established. It will be evaluated whether the structure of the organisation has changed and if so how. It will be shown that during the second reform power stakeholders, such as the former state and collective farm managers continued to manipulate land distribution and that they influenced the creation of participatory, bottom-up organisations. In addition continues the show the linkages between different levels and also indicates how the context in which the organisation is changed, influences the outcome of change.

In both parts of the chapter the theoretical framework of institutional change is utilised to provide explanation of the local level changes.

2 The first land reform

2.1 Background to the reform after independence

With independence in 1991, Uzbekistan had 971 collective farms and 1,137 state farms. (Ilkhamov 1998, p. 539; Bloch 2001, p. 18) Together these farms covered 4,2 million ha of irrigated area. In 1992 and 1993 the first privatisation of land took place. 500,000 ha (around 12 per cent of cultivated land) were distributed among former state and collective farm employees as small household plots. (compare Spoor 1995, p. 51) According to Pomfret the area of smallholdings 'increased from 2.3 million in 1990 to 2.9 million in 1995' (Pomfret 2000, p. 271)

Furthermore, the Law on Peasant Farms stimulated the creation of 9,254 peasant (dekhkan) farms from July 1992. In 1994, 100,000 ha of land was allocated to establish livestock farms. Farms could receive 0.3 to 2.0 ha per head of cattle. (Spoor 1995, p. 51) Pomfret gives similar accounts. However, according to him not new land was distributed but old livestock farms were sold off. He stated 'further decrees in 1994 authorised the creation of private livestock farms (600, mainly unprofitable, livestock farms had been auctioned off by the end of 1995) and private vineyard and orchards.' (Pomfret, 2000 p. 271)

The number of peasant farms doubled within three years. By the end of 1995 the number of peasant farms had increased to 21,514. (compare Pomfret 2000, p. 271) Ilkhamov uses similar figures. He argues that Uzbekistan had in the beginning of 1998 20,000 independent peasant farms. (compare Ilkhamov, 1998 p. 547) According to Ilkhamov in 1997 1,177 of peasant farms ceased operations. (Ilkhamov, 1998 p. 548) But, these figures are contested. Bloch's research uses different figures. He argues that the number of private farms was in 1997 20,700 and in 1998 23,200 in Uzbekistan. (Bloch 2001, 17) Furthermore, while Ilkhamov indicates that there was/is a 'crisis in the development' of the peasant farms, with a 'performance slump' and bankruptcies (Ilkhamov, 1998 p. 548) Bloch shows that the numbers of peasant farms are increasing instead of decreasing. According to him, the number of peasant farms increased from 1997 to 1998 by 2,500. (Bloch 2001, p. 17) The official data from the Ministry of Macroeconomics and Statistics is presented by Kropp. (Kropp et al, forthcoming, p. 9)

Figure 4.1 Change of number of farms

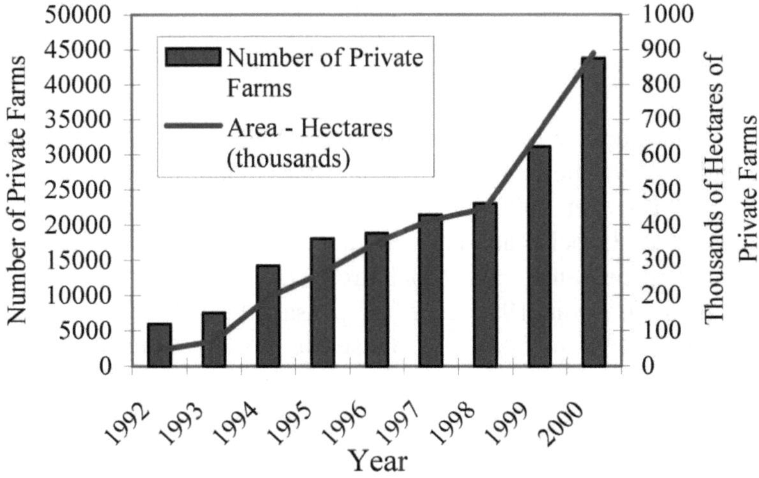

In Uzbekistan land distribution to peasant farmers is on lease basis. Bloch states that 'one specific feature of privatisation is connected with the fact that land cannot be private property but only leased'. (Bloch 2001, p. 16) Ilkhamov points out that the leases, which are given to farmers are on a basis of 10 years only. (Ilkhamov, 1998 p. 550) This makes Pomfret's distinction questionable as he argues that out of the 21,514 private farms '9,346 were independent peasant farms with lifetime inheritable possession of use right, over three quarters of such farms had livestock, and 12,168 were leasehold peasant farms formed on land from collectives.' (Pomfret, 2000 p. 271) Because land distribution is based on leasehold only, Pomfret's distinction between leasehold and lifetime inheritable possession does not sound plausible. However, even Ilkhamov differentiates between two different kinds of peasant farms. He distinguishes not by ownership but by the status of legal entity and as subdivision of the state farms. (compare Ilkhamov 1998, p. 547)

The figures for land allocated to households differs according to the author. Pomfret argues that households had in 1995 372,200 ha or 9 per cent of the total land area. (Pomfret, 2000 p. 271) According to Bloch in 1995 household plots had 602,000 ha total or 499,000 ha arable land. By 1997 the figure of household plots land had increased to 650,000 ha of arable land or 15 per cent of irrigated land reserve. (Bloch 2001, p.17, the same figure is confirmed by Ilkhamov 1998, p. 551) Pomfret offered the following numbers of the farm types and their land area, the data was based on the year 1995.

Table 4.1 Different farms in Uzbekistan 1995

Farm Type	**Number of Farms**	**Area (1000 ha)**	**% of total land area**
State farm	16	152.1	3.7
Collective farm	1,389	1,937.0	46.6
Leasehold dekhkan	12,168	297.2	6.7
Indep. Dekhkan	9,346	230.2	5.5
Smallholdings	2,869,069	372.2	9.0

Source: (Pomfret 2000, p. 271)

The evaluation of the available data indicates that the main emphasis of the government was not on privatisation but 'destatiation' (compare Bloch 2001 p. 15), or the transformation of the state farms into different organisations. The government initiated the reform process and controlled it. In 1993 nearly all of the state farms (95 per cent) were transformed into joint stock companies, co-operatives or collectives. (Ilkhamov, 1998 p. 539) 'A January 1993 decree laid out a program for restructuring the 1,061 state farms and by the end of 1995 only 16 of these remained. Most of the state farms were turned into collective farms.' (Pomfret, 2000 p. 271) These different economic units functioned like collective farms, the government controlled them through the state order system.

By 1996, the number of collective farms had increased from 971 in 1991 to 1,374. (Ilkhamov, 1998 p. 539) The amount of land privatised during the land reforms indicates that privatisation and the establishment of private farms was not the main concern of the Uzbek government.[22] The following section will show that during the first land reforms the power structure did not change, rather it was replicated.

2.2 *The influence of the local elite and district elite on land allocation*

Officially the land reforms were organised fairly and equally. Every farm employee could apply for land from the state and collective farm. Farm employees wanting to be independent farmers had and have to go through a long administrative process. Eckert describes the process in the following order:

> In order to obtain land for the founding of a Dehkhan, the applicant first has to apply to the Economic Council of the Kolkhoz. The Economic Council decides which plot will be handed over to the farmer; the farmer cannot apply for a particular plot of land. The applicant is then given on to the 'village council', which is in fact the kolkhoz council and different from the 'mahalla'. The village council has to approve of the application. The applicant will then address the commune, which sends a surveyor and an 'architect' to the plot. They will survey the land and draw a map, which is registered in the cadastre. Following this, the applicant has to hand in the application complete with the signatures of the economic council, the village council, the surveyor and the architect to the Commission for the Realisation of Economic reforms. The commission will give a recommendation to the Hokhim to grant the application. (Eckert 2000, p. 18)

However, it seems that different sources describe the administrational stages and hierarchy differently. Lerman points out, that

> the village council and the district authorities will not move on a new applicant for land without the consent of the farm manager, who is expected

22 It seems that the main concern of the Uzbek government was to stabilise cotton and rice production and to become food self sufficient in terms of grain production as well.

> to allocate a physical tract of irrigated land to the new peasant farm out of the lands that his farm manages. (Lerman, 1996 p. 163)

In an informal interview, Suleimenov stated that the district government sent the application to the members of the state and collective farm, the board of the members decided on the applicant. However, he added that not all members had to be present and that the board of members made the final decision. The state and collective farm manager appoints all board members. Therefore it would be questionable whether the board would decide against the decisions of the state and collective farm manager. (Informal interview Suleimenov 27.07.2001) Hence, the state and collective farm manager is a mayor actor in the process of land distribution.

According to Eckert the Hokhims 'are nominated by the president'. (Eckert 2000, p. 12) In Uzbekistan, a top-down society with strong Presidential powers, nomination from above implies that there is no alternative. Hence, Renger stated 'the President appoints the Hokims (=govenor) and Deputy-Hokims who are directly subordinated to him.' (Renger 1998, p. 17) Keeping in mind the direct link between President and governor, it can be assumed that in the power hierarchy the state and collective farm manager is below the district government and will take orders from it. Eckert stated the 'kolkhoz chairmen took orders from the Hokhim' and 'The Hokhims had a position of overall control and absolute command'. (Eckert 2000, p. 13) Bektemirov confirmed Eckert's statements pointing out the power position and the influence of the Hokhims over the state and collective farm and the state and collective farm manager. (informal interviews with K. Bektemirov July 2001) This would imply that if the district authority pushed for an applicant to receive land, the state and collective farm manager would not decline the application. Furthermore, it is to assume that because of the pressure from the Hokhim the state and collective farm manager would allocate special plots to the farmer, special in terms of location and productivity.

The former state and collective farm manager is responsible for agricultural land allocation. Keeping in mind that in Uzbekistan 'land is useless without water' (Karimov 1992 in Lerman 1996, p. 155), Suleimenov pointed out, that the state and collective farm manager could allocate land, which has unfavourable agricultural conditions, such as high salinity, water logged area or is at the tail end of the irrigation system. (Informal interview Suleimenov 27.07.2001) Eckert emphasises that farmers 'were receiving land of bad quality. Furthermore, not only did they have to farm the most unproductive land, they also had to travel up to 15 km to reach their plots'. (Eckert 2000, p. 18)

Eckert's argument is confirmed by Egamberdi, who argues that state and collective farm managers 'have also used their clout to push most private farmers onto less desirable, unirrigated lands.' (Egamberdi, p. 241) According to Eckert 'in 75 per cent of the cases Dekhan [peasant farms] land was over 10 km away from the village where the farmer lived'. (Eckert 2000, p. 18) However, it is questionable whether the allocation of 'bad' plots is based on personal decisions of the local and district elite or based on a policy directive from above. Eckert mentions 'a decree of the Cabinet of Ministers about "additional measures for the realisation of economic reforms in agriculture" demands the development of virgin land and the irrigation of bigger areas.' (Eckert 2000, p. 17) Ilkhamov quotes the law 'On the Peasant Farm', 'peasant farms shall be created on the reserve land, special national fund, in the farming lacking labour force, and in the areas of new irrigation' (Ilkhamov 1998, p. 550) Therefore it could be argued that the state and collective farm managers allocating land with unfavourable conditions are only following the requests from the government. How is it possible to show whether the established elite manipulated the land distribution process? If all new applicants received 'bad' plots then this will indicate that every applicant for land got treated equally and that the patronage system did not help to receive better land.

In contradiction with the fair and equal distribution of land, Bektemirov pointed out that land with favourable conditions was reserved for the district government and informally already belonged prior to land distribution to the district government. (Informal interview Bektemirov 27.07.2001) Bektemirov's statement was based on his experience of his village in Karakalpakstan. The village as well as a few other villages were part of the state or collective farm in that area. Hence, the employees of the state or collective farm could not receive this good land. Eckert's research expands Bektemirov's argument on the favouritism to the district government.

Figure 4.2 Power and Influence of the stakeholders during the first land reform

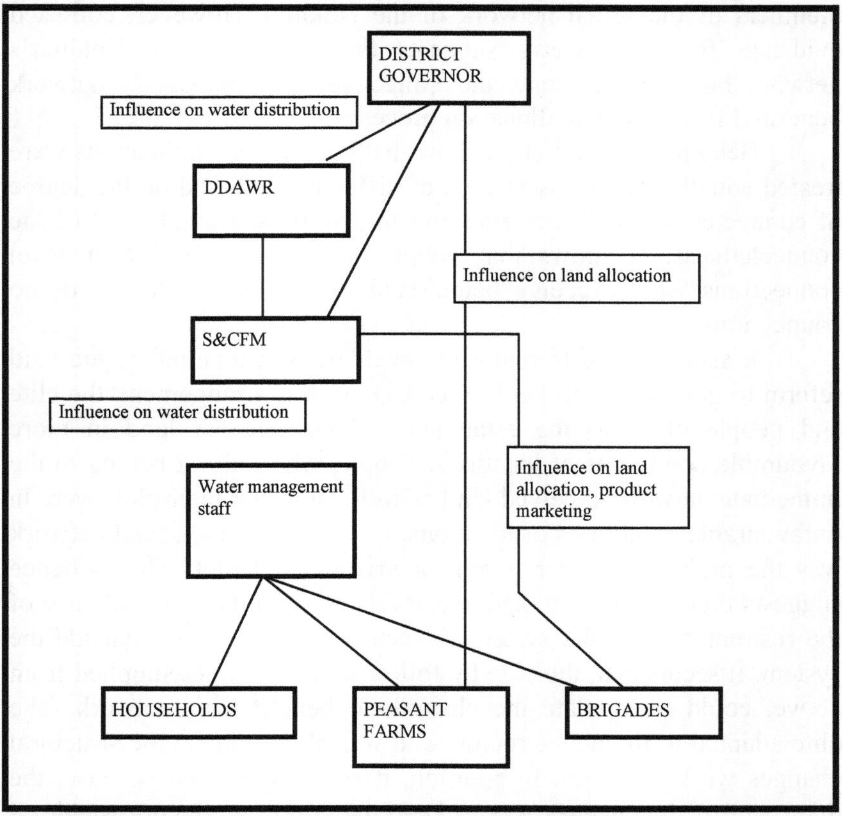

DDAWR: District Department for Agriculture and Water Resources
S&CFM: State and Collective Farm Manager.

Eckert points out that not only the district government received land, but also people close to the district government. She argues that even though new peasant farmers are 'expected to have the experience of a professional farmer [...] In many cases though, members of the administration were the first and sometimes only ones to found a private farm'. (Eckert 2000, p. 20) She continues, 'the personal acquaintance with the Hokhim thus becomes an important factor in obtaining a lease,' (Eckert 2000, p. 20) The beneficials are the members of the social network of the Hokhim. However, collected evidence from interviews suggests that not only the Hokhim's network but also the state and collective farm manager's network benefited from the land allocation process.

Bektemirov and Eckert show that not all land applications were treated equally. Their observation of difference is based on the degree of connectedness with the elite and the administrational level of the connectedness. It shows that people with good and higher level connections would receive better land than others with less or no connections.

It seems that different elite levels tried to manipulate the land reform to gain access to land. According to this arrangement, the elite and people close to the established elites received land in more favourable conditions or locations. People, who did not belong to the immediate network, received land as well. However their plots were in unfavourable conditions or locations. In both cases the social network was the main reason for getting a private land plot. The evidence suggests that the elite on different levels of the structure made use of the restructuring of the system to secure their livelihood outside the system. It seems that, the elite by following the changes supplied from above, could manipulate the changes to benefit their network. The elite adapted to the new structure and was able to make the structural changes work for them. In addition, it could be argued because, the main aim of the changes was to keep the cotton production stable, it did not matter to the top-level, that the changes were not transparent and the implementation was based on the main stakeholders.

2.3 *Habits, institutions, and adaptive capacity*

According to the administrators of the land reform, the number of applications to establish a private farm was rather low. Administrational bodies argued that farmers did not want to become independent because of their lack of experience of independent farming. Constraining factors on the privatisation were not only the bureaucratic hurdles, which made the process of applying difficult, but also the lack of experience in managing and in independent decision making. This lack of independence was one of the reasons that not all peasant farms were successful after becoming independent. Ilkhamov argued that 'the decades of the kolkhoz-sovkhoz system were bound to leave a heritage; they helped to form a type of toiler with an inhibited sense of ownership and undeveloped economic thinking' (Ilkhamov 1998, p. 549) Ilkhamov reasoning reflected the dominant explanation for the small numbers of the new peasant farms. Hence, 'many people did not want land because of the socialist mentality, which shrank from self-reliance and initiative' (Eckert 2000, p. 16) According to this reasoning individuals had a low adaptive capacity to adjust to the changes.

However, to mention the 'socialist mentality' as the main reason for the small numbers of independent farm establishments could be wrong and just a method of simplification so that other issues would not be addressed. Here 'socialist mentality' could be defined as habits, which are according to Bourdieu difficult to change. However, it could be reasoned that the 'socialist mentality' was contradicted through the private sector household plots, which worked very successfully during the Soviet period. However, as reasoned by Eckert 'Private yards do not demand the same level of management skills, marketing opportunities and technical expertise as full time private farming.' According to her, farmers lacked exactly these skills. Eckert argues that farmers calculated their benefits and reacted risk adverse. This would imply that not a 'socialist mentality' prevented them from applying, but a calculation of the skills required. Eckert points out that during interviews farmers expressed 'a lack of experience in management, accounting, generally in small scale agriculture and specific agricultural questions which formerly were delegated to specialists employed by the kolkhozes and sovkhozes'. (Eckert 2000, p. 16) She argues, that the farmers were aware of their deficit and expressed 'a great interest in overcoming it through training.' (Eckert 2000, p. 16) However, her argument contradicts some of her own findings. She reasoned that members of the administration applied for land and received it. The administrational staff had probably the same level of or even less agricultural skills and technical expertise as the farm employees. Hence, it seems that her argument of lack of experience leading to lower application cannot be the main reason for not applying for land.

There are at least two ways of reasoning against an institutionalised 'socialist mentality', firstly it would be possible to argue that for different actors an institution is differently manifested. Hence, for some a 'socialist mentality' would be less a hindering factor than for others. Taking the evidence, that most farm employees and administrators with connections applied, this could imply that for them the 'socialist mentality' was not evident. However, this reasoning implies that there is only one institution or the other, either 'socialist mentality' or independent/economic thinking. This reasoning seems to be limiting, because it implies that there is only one institution influencing the actors, however as it became explicit in the background chapter there are more levels of influence continuously affecting the actors. According to this reasoning, it is questionable whether the 'socialist mentality' was the most dominant 'mentality' affecting those applying for land, or whether the 'patronage mentality' was more dominant, and therefore even actors with a 'socialist mentality' applied because they thought they could rely on the patronage system for help. Hence, different structures influenced the perception of the actors, and the actors chose the options which looked most promising to them with their restricted institutional view.

It is possible to find a compromise between the two strings of reasoning – there is more then one institution and these institutions are differently manifested. Hence, not every actor who had connections applied for land, but only some. Others choose different options, because these other options looked more promising. In terms of adaptive capacity, it shows that the more institutions are available, the more choices the actors have, hence the more adaptive they can be.

2.4 *Outside influences, risk and adaptive capacity*

It seems that the transaction costs of applying for land are lower for an actor with connections. In addition, it might be possible to argue that the level of outside risk for an actor with connections is also smaller.

The surroundings of the land reform were not in favourable conditions for independent farmers. Lerman points out that the market services and infrastructure are not functioning. (Lerman 1996, p. 163) Ilkhamov describes the conditions of the early land reform, as 'prices for power, fuel lubricants, fertilisers and equipment are sky-rocking and credits cannot be obtained' (Ilkhamov 1998, p. 549) It is questionable whether new land owners with connections and without connections had to face the same unfavourable conditions. Even though, it is not possible to make sweeping statements, one has to consider that networks might soften the wider conditions and therefore have allowed more flexibility and dynamics for the actor. If the structure is static or seems to be static, actors might not want to challenge it.

According to Eckert it seemed that the farm employees were aware of the institutions and assumed that information and training was one way to adapt to the institutions or at least to make them more flexible. On the other hand, training and special knowledge does not help with the institution of patronage. However, they do help in relating to the institutions which utilise 'scientific knowledge', which was emphasised in the later stages of the Soviet Union. Hence, here the actors are responding to a different institution, which they identified to be the most helpful and promising for their situation. Eckert argues that farmers admitting their lack of knowledge and experience is 'a realistic estimation of the farmers side of their own abilities and knowledge.' (Eckert 2000, p. 16) On the other hand, maybe it was realistic within one institutional perspective, but not within a different institutional perspective.

Eckert reasons that, to establish a private farm, farmers had to take a higher risk, then to comply to the ongoing changes and to continue working on the collective farms. Eckert supports her reasoning by arguing that farmers saw the brigades 'as a low risk opportunity to train for future private farming'. (Eckert 2000, p. 16) However, the practice shows that the brigades did not differ from the already established units at the state and collective farms (compare Ilkhamov 1998, p. 544) therefore it is questionable whether farmers could use the experience of the brigades to train for future private farming. This low risk opportunity seems more like a second best option, because other alternatives were not available, and the perceived institutions seemed too restrictive, hence the actors continued with what they knew best.

Douglas argued that in general institutions do not change from the inside, hence actors replicate the current institutions, analysing the data available this argument seems to hold true. However, it seems that her approach and participatory rural appraisal (PRA) methods share similar problems. Mosse argues that in participatory rural appraisal exercises only the people present in the exercise are considered, but people who do not take part in the exercise have no voice. One could argue that the people who left, chose a different path. On the one hand one could argue that they broke with the existing institutions on the other hand this was just a different available institution, if all the others were too restrictive.

2.5 Power beyond land allocationi to independent farmers

The state and collective farm managers were not only empowered to allocate land to the new independent farmers. The new private farms relied also on the farm manager for 'input supply, product marketing and even lobbying at the district government level'. (Lerman 1996, p.

163) Pomfret and Ilkhamov confirmed this view. Pomfret argued 'The collective often control other inputs, such as fertilisers and sometimes the marketing of outputs. Thus, when the collectives' administrators 'recommended' to peasant farmers what they should produce, their advice was heeded.' (Pomfret, 2000 p. 272). Ilkhamov stated, that peasant farmers were completely dependent on state and collective farm organisations for economic and administrative services. The dependency of the peasant farmers on the former organisational hierarchy was also reflected in the attitude of the former state and collective farm managers to the newly independent peasants.

Ilkhamov argued that 'many kolkhoz chairmen [state and collective farm managers] consider peasant farmers as their own subdivisions and dictate to them what to sow and to whom to sell their products.' (Ilkhamov, p. 550) While Lerman called the dependency of peasant farmers to the former state and collective farm manager a 'friendly' arrangement, Pomfret and Ilkhamov were more negative about the relationship between independent farmer and manager. Ilkhamov used the term 'subdivision' and 'dictating'. However, it is questionable whether state and collective farm managers would treat newly independent farmers with connections to levels beyond the state and collective farm negatively. Because it would be possible for these farmers to address bad treatment to higher levels, and therefore destabilise the position of the state and collective farm managers. This would imply that the first land reform established first and second class farmers, namely those with and those without connections.

Overall, even though officially land reforms established the possibility of independent farming, in reality the formal organisations did not change. Hence, the patron-client relationships between state and collective farm and former farm employees continued as well. One has to keep in mind that, the state order system for crops was still in place hence land reforms did not imply independence. The state and collective farm managers would have been responsible for the production of these crops on the former state and collective farm land. Hence, the state and collective farm managers would have risked their own position, if they did not fulfil the target, to sustain their power position they had to follow the orders from above. Taking into consideration that some farmers had connections to higher levels, it seems that the position of the state and collective farm managers became overall more unstable. This could have increased the pressure on the new farmers without connections.

In addition, the dependency of the new independent farmers on the former state and collective farm managers was reinforced through the system of water distribution. The district water institution was responsible to deliver water to the border of the state and collective farm. The state and collective farm was responsible for the on-farm irrigation system, 'that is: the distribution of water and the maintenance of the irrigation canals on its land'. Hence, 'Private farmers have so far no individual contracts with wotkhoz (district water institute) but depend on the kolkhoz.' (Eckert 2000, p. 31) This implies that they depended on the hydrotechnical units of the state and collective farms, which controlled the distribution of water. The state and collective farm manager appointed the head hydro-technician. Eckert points out, 'there were farmers who refrained from applying for a lease because they said it was too difficult to obtain water from the kolkhoz or directly from the wotkhoz.' (Eckert 2000, p. 32)

Egamberdi stated that even though 'on-farm irrigation management is supposed to be the responsibility of the farmer [...], the state and collective farms have retained responsibility for distributing water to dehqan and private farms'. (Egamberdi, p. 240–1) Egamberdi's statement has to be viewed in relative terms, for two reasons, firstly the state and collective farms still exists, and secondly in the more recently established irrigation systems (see background chapter) water was delivered to the border of the state and collective farm. This implies that the 'great majority of irrigated lands belong to collective farms, which utilize most of the water.' (Egamberdi, p. 240–1) and that the state and collective farms continue to be responsible for water distribution. However, it could be questioned whether the water distribution among the new and old users are equal and fair. The hydrotechnical units of the state and collective farms could use water as a tool to achieve compliance. Egamberdi confirmed this, arguing that state and collective farms 'take advantage of their control over the current shortage of water. They use their distribution power to impose their will on dehqan and household farmers by withholding water. (Egamberdi, p. 241)

In recognition of the contribution of private households towards agricultural production, the state allocated larger land plots to households with the purpose of securing the food self-sufficiency of the rural population. Nevertheless, households were still dependent on state and collective farms. Households had two different types of land, the first was the plot which was allocated from the local government as living space. The second was the plot that was distributed from the state and collective farm as land for agricultural production. Article 21 of the Law on Land stated that the size of the allocation of land depends on the 'availability of land, participation of the recipient in the work of the collective farm, the limits set by the enterprises charter, and the opinion of the administrator', (Basis Report, Annex D) The state and collective farm manager could decide which agricultural land would be allocated to the household, so it was up to the state and collective farm manager to decide the size of the plot and whether the plot was in a more favourable location, close to the village or very distant. Therefore, the establishment of a household plot was under the direct control of the state and collective farm. After land allocation, the households were dependent on water allocation from the state and collective farms, here the dependency is similar as to the independent farms.

2.6 Summary of the first land reform

The literature review of the first land reform indicates that the elite within organisations, which were responsible for the implementation of the organisational reforms, could manipulate the anticipated changes from the top to their own benefit. The reasoning of Egamberdi, Pomfret and Ilkhamov indicate that the state and collective farm managers kept their power positions. Spoor is more vague about who gained, he called it the nomenclatura; however Eckert distinguished between district governor and state and collective farm managers. It would be possible to argue, as already indicated above, that because of the connections necessary to gain access to land, that the top-down power structure on the local level is weaker than before the reforms. Then the local farm employees had less possibility to access higher levels for support, with the entrance of members of the administration from higher levels, the power of the state and collective farm managers decreased. This would imply that some form of transparency towards the lower levels, or at least some parts of the lower levels, occurred. According to this reasoning, the state and collective farm managers would have a decrease of security after the reforms, because of the direct linkage of some new farmers with higher levels. This reasoning questions Pomfret and Ilkhamov statements of a top-down order, and argues that the situation became more complex with more linkages between different institutional levels after land reforms.

The state and collective farm managers continue to be the gate keepers and facilitators of the local community. However, other higher levels became direct gate keepers and facilitators for the local level as well. These higher levels have influence on the state and collective farm managers. While Eckert argues that the reform reduced their top-down risk situation, here it would mean the responsibility to fulfil the production plan, now smaller brigades and the new independent farmers became more responsible. On the other hand, reforms increased the top-down risk, because of the linkages between independent farmers to the district or even the provincial level. This would also imply that the bottom-up risk, of criticism and non-compliance and therefore reduction of social status increased.

3 The second land reform

3.1 Background to the establishment of water user associations in Uzbekistan

In 1996 the government of Uzbekistan contracted the Central Asian Irrigation Research Institute (SANIIRI) to research the experience of other countries with water user associations and to establish a framework for the legal and organisational establishment of water user associations in Uzbekistan. In 1998 the European Technical Assistance to the Commonwealth of Independent States (TACIS) started a project in co-operation with the government of Uzbekistan, which included 14 private farms on parts of the territory of a former state and collective farm. The objective of the project was to rehabilitate the irrigation system, to train the farmers and to establish the first water user association.
The Uzbek government could not contribute the promised governmental funds, therefore TACIS could only finish its own part of the project (informal interview with Dr. T. Bekmuratov, expert of SANIIRI on water user associations, July 2000). Consequently, the rehabilitation of the irrigation and drainage system remained unfinished within the project area. In June 2000, the control of the system in the pilot project was handed over to the farmers and since then the water user association has been working independently.

In the beginning of 1999 SANIIRI completed its work and presented the results to the Ministry of Agriculture and Water. In February 2000 the government decided to continue with state and collective farms privatisation. Unproductive state and collective farms became privatised and officially their farmlands were distributed among the former farm employees.

Since Uzbekistan's independence, and the stopping of all financing from Moscow followed by the economic collapse, the maintenance of the irrigation system declined. According to Renger, since 1995 the state and collective farms were responsible for the maintenance of the on-farm canals, irrigation networks, the drainage systems and for measures against the rising level of salinity. (Renger 1998, p. 17) However, the reform did not have any impact and the quality of the irrigation systems declined further. The government assumed that privatisation would deepen the gap in fulfilling the required service and ordered by decree the establishment of water user associations on privatised state and collective farms. It was intended that water user associations would take over the former state and collective farms' role as a service provider for the irrigation system.

The government originally intended to rehabilitate the old on-farm irrigation system fully and to hand over a restored and perfectly working system to the newly formed water user associations. It was also anticipated that the farmers would be trained and prepared for becoming independent farmers. By summer 2000, twelve water user associations had been established in three provinces of Uzbekistan, five in Khorezm, five in Karakalpakstan and two in Syrdarya, with respectively a total area of 9.500 ha, 13.500 ha and 6.000 ha. In addition, Syrdarya had two unregistered water user associations. (Informal interview with Dr R. Ikramov, General Director of SANIIRI) Rehabilitation work on the established water user associations did not take place, with the exception of the Syr Darya TACIS pilot project.

3.2 Location of the evaluated water users associations

The water user associations visited and analysed were in three different provinces, Karakalpakstan, Khorezm and Syr Dar'ya. The water user association in the Syr Darya province is the water user association of the TACIS pilot project, meanwhile a second TACIS water user association pilot project has started. (informal interview with Wob Groot Consultant on water user associations of the second TACIS project) In Khorezm two water user associations have been visited, one in Khiva and one in the Urgench district. After the research conducted in 2001, the water user association in Urgench has been chosen for future work together with the United States Agency for International Development (USAID) Natural Resource Management Project (NRMP). Two water user associations have been visited in Karakalpakstan, one in a Southern district, Beruni district, and one in a Northern district, Keigili district. While the water user associations in the Syr Darya province and in the Khiva district of Khorezm were visited in August 2000, the water user associations in Urgench, Beruni and Keigili were visited in July 2001.

3.3 Sizes of the water user association member farm

The water user association in the Syr Dar'ya province consisted of 14 farms and 16 households. The average size of the farms was 15–30 ha and the size of household plots approximately 20 ha. The situation in the Syr Darya province water user association is an exception, its uniqueness is due to the fact that its establishment is based on the TACIS pilot project. The other water user associations differed in size from the water user association in the Syr Dar'ya province. In Khorezm, the water user association in Khiva district had 53 private farms with an average of 14 ha each and the water user association in Urgench had 68 private farms, their size varying between circa 10–100 ha. In Karakalpakstan, the water user association in Beruni had 74 and in Keigili 88 private farms, the size of the farms was similar to the water user association in Urgench. The size of farms varied with the specialization of the private farmer. In the water user association in Urgench were 11 farmers specialized in fruits, 1 in cattle and 56 in cotton production. The size of the farms specializing in cattle was 97 ha while the 11 farmers specializing in fruits shared only 46 ha altogether and the 56 cotton farmers shared 1430 ha, with an average land size of 25 ha.

The water user association in the Syr Dar'ya province had two distinct types of members: private farms and households. The difference between the farmers and household was not based on the size of the land plot, but on the length of leasehold ownership.

Farmers had land on lease for 10 years while households had only a 1-year lease. While in the Syr Dar'ya province households and farmers could become members of the water user association, this was not the case in Karakalpakstan and Khorezm province. There, the households, which had small plots of irrigable land, could not become members of the water user association. In Karakalpakstan and Khorezm province the number of households was much higher and the size of the household plots smaller in comparison to the number and size of households and their plots in the Syr Dar'ya province. The water user association in Urgench had 870 households, which had in total 95 ha of land. Hence, the average land allocation was only 0.1 ha.

While the size of the water user associations in Karakalpakstan and Khorezm were based on the size of the former state and collective farms, which had been privatised, the water user association in the Syr Dar'ya province was only part of one state and collective farm. These differences were reflected in the variation in size, participation and organization of the water user associations. For example, while the water user association in the Syr Darya province consisted of only one village, the other water user associations consisted of more villages with an average of approximately 1000 households. (water user association in Beruni district had 1150, Kegeili 1400 and Urgench 870 households.)

3.4 Land allocation in the second reform

Even though the process of land distribution did not change land should have been distributed on a competitive basis. According to Lerman 'the applicants 'bid' for land by submitting their business plans and stating their qualifications' (Lerman 1996, p. 159). The process of application with a business plan was confirmed by SANIIRI officials in Tashkent (informal interview with Dr. T. Bekmuratov, Tashkent 2000) and by the province manager of water distribution in Khorezm, (informal interview with B Babajanov, Urgench 2000). However, evidence from field visits indicates that the competitive process of application is not the main reason for land allocation. A field visit to the water user association in Syr Darya province and Khiva district showed that former state and collective farm employees without farming experience received land. In Syr Dar'ya one independent farmer gave up his land after one year. During an interview with the other farmers, it was stated that he was the former cook of the state and collective farm and a relative of the district governor. In Khiva a former tractor driver of a collective farm received land. It is questionable to what extent former specialised employees of the state and collective farm were able to make a realistic 'business plan', therefore, it is questionable whether the land reform was 'competitive' or whether land allocations was due to informal and formal networks. It seemed that the same networks responsible for land distribution during the first land reform were influential during the second land reform. Spoor stated that 'with the system of allotting land through the powerful khokims or mayors, vested private interests in the public sector promote land to the benefit of the nomenclatura' (Spoor 1995, p. 53).

As mentioned for the first reform, the influence of the district governors played a major role in the distribution of land this – was similar during the second reform as well. In the water user association in Urgench district a former employee of the province water distribution institute received agricultural land to start a farm. In an informal interview he stated that he applied for land for two reasons, firstly he wanted to be more independent and secondly he had health problems and wanted to spend more time at home. His family was running the farm, while he saw himself as the manager. However, according to the regulations, he would have not been eligible to receive land, because he was not a farm employee and because of his health problems. The land reform allowed old stakeholders to receive land. Suleimenov in the Kungrad district in Karakalpakstan stated that people with networks and money, receive land in more favourable locations. (informal interview with Suleimenov June 2001) Also, in the water user association of the Urgench district, the farm organisation manager stated that two former collective farm managers from other farms were now newly independent farmers in the water user association. It was stated, that these two had informal networks to higher administrational levels, and utilised these networks to receive land. (informal interview with Urgench Farm Organisation manager June 2001).

Bektemirov argued that prior land reforms, land in favourable conditions belonged informally to the governors, the land reform formalised the informal arrangements and created security for the people in power positions. (Informal interview Bektemirov 27.07. 2001) An example from the Kegeili district indicates how the former state and collective farm manager took over land in favourable positions in terms of access to water. The field visit to the water user association in Kegeili showed that the state and collective farm manager was one of the few farmers who had access to water during the drought in 2001. While he had access to the main irrigation canal the other farmers had to depend on water from the secondary canal, which had no water. The new institutions allowed the old stake-holders, who where the agents of change, to formalise the informal claims to agricultural land and use land allocation as a tool in the patron-client relationship.

3.5 Water user associations and farm organisations

The visited water user associations were integrated into a larger organisational framework, the farm organisation. A farm organisation was established on every former state and collective farm and took over the duties of the former state and collective farm. Eckert states 'the associations of Dehkhan farmers were founded by RAPO and the Hokhimiats apparently on demand of the farmers in order to advise them.' (Eckert 2000, p. 13) However, Pomfret argues, that these organisations 'were introduced on a top-down basis and were not regarded by peasant farmers as their organisations'. (Pomfret, 2000 p. 281 footnote 9) Eckert states that the farm organisations 'are headed by members of the local administration nominated by the Hokhim'.

(Eckert 2000, p. 13) Her statement confirms Pomfret's argument of top-down introduction. However, it does not verify whether the farmers thought of the organisation as 'their' organisation.

If there had been fair elections, arguably the positions of former state and collective farm managers would have helped them to gain the positions as farm organisation managers. The reasons for their election were numerous: such as familiarity with the people, experience of management positions and established connections to other institutions. A competitor, as a new bottom-up person, would have had high transaction costs to gain a similar level of appeal in the election, amongst the local level; furthermore, his/her networks to the established higher levels could be weak, hence the communal benefits would be lesser than with a person with strong networks within the hierarchy. In addition, experiences of water user association in Kyrgyzstan indicate, that farmers delegated any form of responsibility to former power holders. (Baumann 1999, Johnson III 1999) An explanation for this behaviour is that the discourse and habit rules have not adjusted to the new situation and therefore contribute to the election of the former managers. A combination of these arguments indicates that even in fair elections the former managers may have gotten the job anyway.

Eckert argued that the farm organisations

> are supposed to (a) advise the private farmers on farming questions, e.g. when and what should be planted; and (b) to assist them in organising transport and marketing, e.g. procurement of the contracts with the canning factory, the grain mill or the cotton plant. (Eckert 2000, p. 13)

However, giving advice and assistance to the farmers does not necessarily change the farmers' perception of the organisational structure and the way the organisation was introduced. However, it might build trust between the organisation and the farmers in the long-term. Pomfret stated that farmers did not perceive the farm organisations as organisations that 'could be mobilised in support of their interests'. (Pomfret, 2000 p. 281 footnote 9) Hence, the organisations were still perceived as reflections of the old style top-down organisations. Furthermore, as stated above the former management 'ordered' the farmers what to produce and where to market their products. If the same people would advise, in effect give the same advice as the former order, then the farmers did not see the difference between state and collective farm managers and farm organisation manager. In addition the system of the state order system for rice, grain and cotton still existed, this implies also that the advice in terms of production and where to market was still a top-down order. Furthermore, because the farmers did not have the chance to get advice on agricultural production from alternative sources, they did not have the means to compare and to judge on the assistance.

Even though, farm organisations were and are supposed to advise and assist farmers, which are all members of water user associations, in Karakalpakstan at the time of the field visit in July 2001 none of the members of the farm organisations or water user associations received training. This confirmed earlier experience from fieldtrips to the Khiva water user association in the year 2000. The field experience showed that even though the farm organisations were introduced officially with the intention to benefit the farmers, these benefits did not reach the farmers. However, training and assistance could also be interpreted as a function to keep the farmers satisfied with the system, so that they would not try to change it. The evidence suggests that even in the Urgench water user association in which training took place, farmers did not question the structure of the organisation. Hence, the training came within the organisation and kept the farmers satisfied with the organisation.

3.6 *Top-down or bottom-up strategies for initiating a water user association*

As stated above, the formation of water user associations was ordered from the national level. The state initiated the research in water user associations and contracted SANIIRI to provide guidelines for water user associations to be set up and to give advice to the new organisations in Karakalpakstan and Khorezm. According to SANIIRI experts from Tashkent, SANIIRI not only worked very closely with the established water user associations, but also established research stations in the villages to keep record of the changes. However, while the field observation in Syr Dar'ya and Khorezm province confirmed the close relationship between SANIIRI and water user associations establishment, in Karakalpakstan SANIIRI was not involved in water user associations establishment and monitoring. (interview with R. Koshekov July 2001)

Furthermore, Koshekov stated that it was not required to establish water user associations on the former state and collective farm territory, this is in direct contradiction to the statements given by SANIIRI in Tashkent. Koshekov continued that farmers in Karakalpakstan demanded the establishment of water user associations on the privatised farm territories. However, fieldtrips to former state and collective farms show that only the former state and collective farm managers as farm organisation managers received information and the guidelines about water user associations and that the new farmers were not aware of water user associations prior to the second land reform. The former state and collective farm managers also had to inform the farmers about their duties and their responsibilities as members of the water user associations. Hence, it was the responsibility of the former state and collective farm managers to distribute information and to organise water user association elections. The initiative to establish water user associations in Karakalpakstan came from above, and was not a bottom-up movement.

The Karakalpakstan experience confirmed previous fieldwork results (2000) from the water user association in Khiva. At that time, the former state and collective farm manager had the responsibility to establish the water user association. However, interviews showed that farmers were unaware of the established water user association. An interview with a farmer, who was identified prior to the interview as member of the Khiva water user association, indicated that so-called water user association members were not even aware of the water user association. Hence, it seemed that the establishment of water user association was more a formality, which had for the farmers no direct consequences. (see below)

The fact that representatives of the state or the former state and collective farm had to initiate water user associations, indicates that water user associations are not established from the bottom-up. That water user associations were established from above and within a hierarchical structure. They were not local movements or initiatives. Even though the formation of the water user association in Syr Dar'ya province differed, here the former state and collective farm manager was not in charge of the formation process but TACIS. But, even here the initiative for establishing the water user association was not from below, but from above and the independent farmers followed.

Even though, the information for water user associations was distributed from above, it would be possible to argue that farmers wanted to establish water user associations when they were aware of this form of organisation. However, interviews with the SANIIRI officials indicate that farmers did not have any other choice of becoming a member of the water user associations. This was confirmed during interviews with water user association board members. The question, how farmers, which are not in the water user association, are treated, was not understood, because non-membership was not an option. Consequently, there was no special treatment or need for arrangements for non-members. None of the water user associations visited had farmers within the territory of the former state and collective farm, who were not members of the water user association. That none of the farmers were resenting the new organisation, could be explained, either because they did not have a choice, or that they were not informed about the water user association and just accepted the water user association as prerequisite for establishing a farm. However, according to SANIIRI farmers were interested in establishing water user associations.

3.7 Interest in establishing a water user association

During interviews with SANIIRI experts in the year 2000, it was found that farmers were happy that the old system was disappearing and that farmers were willing to establish water user associations. As confirmation of this trend was mentioned the increase of members in the water user association in Syr Dar'ya province, the number of members increased from 14 to 30. However, prior to the formation of the water user association in the Syr Dar'ya province TACIS had partly repaired the irrigation system for the first members of the water user association. It could be assumed that the late joiners were hoping that their irrigation systems would be rehabilitated as well. Hence, the late joining farmers had an interest to participate in the water user association in Syr Dar'ya because the first farmers under the TACIS pilot project had their irrigation system repaired. However, this did not mean that they were concerned with the regulation of water distribution or the operation and maintenance of the system.

During an interview with the farmers of the water user association in the Syr Dar'ya province, it became evident that the water user association was not active. No meetings had been held and operation and maintenance work on the irrigation and drainage system had not taken place since the departure of TACIS. This shows that the main aim of the late joiners was not necessarily the concern about water distribution, but might have been linked to the financial investment of TACIS into the irrigation system. Groot confirmed that the interest in the water user association was not high, or at least not an immediate issue for the local farmers, by stating, that as soon as TACIS left after establishing the pilot water user association, the water user association stopped functioning (informal interview Bishkek July 2001).

Even though the TACIS pilot project failed in the eyes of TACIS, SANIIRI experts anticipated that the number of water user associations would rise in future. Koshekov in Karakalpakstan and Babajanov in Khorezm confirmed SANIIRI's predictions. Koshekov expected that the number of water user associations in Karakalpakstan would double and Babajanov thought the number would increase by four new water user association in Khorezm within the next year. However, as stated above, one of the conditions for privatisation was the establishment of water user associations on the territory of the former state and collective farm. Hence, the increase in water user associations would be forced and not be based on a local demand. The local level was not interested in the changes or did not think that the changes were significant.

3.8 From state farm hydrotechnician to head of the water user association

The introduction of the farm organisations did not change the former structure of the state and collective farm but reconfirmed the old structure. This was also the case for the water user associations. While at the state and collective farm the hydro-technician was responsible for water distribution the head of the water user association took over these responsibilities. Berkoff states state and collective farm managers 'are responsible for water distribution and surface drainage within the collective farm' (Berkoff 1994, p. 3), he specifies that state and collective farms 'employ between 1–4 water management staff for water distribution' (Berkoff 1994, p. 5). While in the early stages of fieldwork it seemed that the former state and collective farm manager became head of the water user association. (observed in Khiva 2000) The early findings could not be validated during the second fieldtrip in the water user associations in Karakalpakstan and Urgench district. During a meeting with members of the Urgench water user association, it was stated that the farmers voted for the old hydro-technician as new head of the water user associations. This was also the case in

Beruni district. However, in the water user association in Kegeili district the former state and collective farm manager appointed the head of the water user association. Strikingly, in all the water user associations visited, the former head of the hydro-technical unit was the first head of the water user association. As indicated above the hydro-technician were former employees of the state and collective farm, and directly subordinated to the state and collective farm manager.

The only exception was the water user association in Urgench district. Here the old hydro-technician was elected as a head of the water user association. After a short period, he resigned because of health reasons. The second head of the water user associations was again an engineer, who was working in the hydro-technical unit of the farm. According to SANIIRI the head of water user association did not have to be a farmer or a specialist. This information was also known to the farmers electing the head of the water user association. In the Beruni water user association, the farmers pointed out in an official meeting that they were aware of the fact that the water user association manager did not have to be a specialist. However, farmers wanted to vote for the person they could trust most. The farmers argued that they trusted the former engineer of the state and collective farm, because of his long experience and his familiarity with the people. However, it was an official meeting, with only a few farmers present. It can be assumed that these farmers were especially invited by the farm organisation manager, hence may have belonged to an influential informal structure. Furthermore, the head of the farm organisation was present, it is questionable whether the answers reflected the general perception of the farmers or were answers, which were officially wanted.

Arguably, the familiarity with the people was a major factor, which caused members to vote for the former hydro technician. The fact that the members voted in the second election again for an engineer, who was less well-known, indicates that the familiarity factor was not the only reason for the election. A possible explanation could be the familiarity with the organisation, delegation to specialists or familiarity with the people in power, who recommend him/her. Furthermore, as pointed out in the theoretical chapter elites have legitimacy through expertise and professionalism, in this case the farmers perceived the responsibility of water distribution best served by hydro-technician. The structure did not change, because it was perceived as the best man for the job, it is questionable whether the farmers really knew what the job 'water user association manager' was really about.

The establishment of water user associations was ordered by decree. This initiative from above could imply that the farmers were not really interested in the forming of the water user association. Furthermore it is questionable whether the farmers have been informed prior of the establishment of the water user association about the role and duties of the water user association. All these doubts lead to the suggestion that the farmers were not interested in the election process. Furthermore, that the farmers could not decide about the constitution of the water user association indicates that the organisation was constructed from above, without consideration of farmer interests at the grassroots. Field visits and informal interviews with farmers indicated that farmers did not receive any training on their rights and duties in the water user association. This shows that the farmers did not make an informed decision when they participated in the election process. Furthermore, that the farmers had not the possibility to vote against the establishment of a water user association alienates the farmers from the start and indicates the coercive power from above.

The implementation of water user associations had large impacts on farmers in terms of responsibility for the operation and maintenance for the irrigation and drainage system. With the establishment of the water user associations farmers had to maintain their irrigation systems themselves and it was anticipated that the farmers would have to pay the real price for water delivery. However, the interviewed farmers either were not aware of the current situation or argued, when they were aware of the changes that they would not be able to afford to pay for the operation and maintenance as well as the real price for water. These farmers assumed that the state would not implement the anticipated changes and would not make them responsible for the costs.

3.9 Training of water user association members

During the first fieldwork in 2000, in the water user association in Syr Darya province as well as the water user association in Khiva it was claimed that initial training sessions for the farmers had taken place. However, during interviews it became evident that not all farmers had participated in the training sessions. There was no training given to farmers who joined the water user association at a later date. The experience was similar in Khiva. Here, even though it was claimed that training had taken place, in an interview with a farmer it was stated that no training was given. In addition, in 2000 during a meeting in Khiva it was stated that the farmers did not need training as such. And that the new farmers are already agricultural experts by living on the land and working the land. However O'Hara stated that

> most farmers have spent their lives as employees of large farms and were allocated specific tasks. Consequently, they took no part in irrigation and even when they did, decisions on when and how much to irrigate land were often made by others or by committee. (O'Hara 2000, p. 368)

The visit to Karakalpakstan in 2001 showed that no training was given to the farmers. When asked, farmers were not informed about the duties of the members of a water user association. The answers were vague and general such as payment of fees and water use efficiency in the Beruni water user association, and deepening and cleaning of canals in the Kigeili water user association. However, the farmers were unable to give further detailed information. In an informal interview with the district manager of water distribution it was stated that farmers pay informally the hydro-technician for irrigation work on their fields.

The water user associations in Urgench could provide documents, which indicated that water user association members had received training organised through the farm organisation. During a meeting with the manager of the farm organisation it was mentioned that training sessions for farmers were organised once a year during the winter. During the seminar, which was supposed to last one week the farmers received different information regarding farming and farm management. However, in Urgench the plan for training was not organised by the farm organisation manager or water user association manager, but by a higher level, by the province manager of water distribution.

3.10 Motivation and strategies to manipulate the risk situation

In her article Eckert gives the following example:

> A farmer applied for land to found a Dehkhan. The hokhim and the kolchos administration denied him the land. The farmer felt unjustly treated and wrote a letter to the president. This letter was handed on to the responsible department of the agricultural ministry. They investigated the case and decided that the land had unjustly been denied to the farmer. The farmer was given land. The Hokhim and the kolchos president lost their positions. (Eckert 2000, p. 8)

Even though, it seems that there is one success story of bottom-up complaint, it is questionable whether this story has relevance to the behaviour of other district governors and state and collective farm managers. The evidence from the literature and from the field visits indicates that this case was an exception. However, the example shows that the situation for former state and collective farm managers in Uzbekistan is not secure. Because of the political decision making process the local level decision makers are at high risk of losing their position and influence. Eckert argues, because of the insecurity 'the local administration will try to avoid taking wrong decisions. It will maybe try to avoid decision making altogether and just follow the orders from above'. (Eckert 2000, p. 9) According to her reasoning the change from state and collective farm to independent farmers seem to increase the security of the state and collective farm manager, because the shift reduces his/her responsibility of the management of the whole state and collective farm. This could be especially the case for the established water user associations, all of them were formed on unproductive state and collective farms. However, Eckert pointed out in her research that new farmers were mainly former administrative staff, this would be in direct contradiction with her statement of avoidance of 'taking wrong decisions'.

Before the reform the state and collective farm managers were responsible that the state and collective farms achieved the set production target. Privatisation implies that the independent farmers themselves are responsible for the targeted production for their plots. Furthermore, because the local elites were able to gain the land in more favourable conditions, they might have the possibility to achieve the targeted amount of crop on their own land.

Eckert focusing on the first land reform analyses the interest of the kholchos (state and collective farm managers) and the hokimiat (governors) in establishing either shareholdings or shirkats. She argues:

> The Hokhimiats currently seem to favour the shirkat-system while the kolkhoz administration prefers the shareholdings [...] This is understandable if you consider that within the shareholdings the positions of chairman of the kolkhoz is hardly changed. His powers might even be expanded as he will be more independent from the state but still with equal powers within the Kolkhoz [...] Likewise the Hokhims might prefer the Shirkat system since it resembles the directives by the ministry of agriculture more closely. (Eckert 2000, p. 17)

Analysing the second reform, which only introduced independent farmers, it seems that both the governors as well as the state and collective farm managers could be in favour of the changes. On the one hand the shifts from state and collective farm to privatised farmers on former unproductive farms are ordered from above. This would coincide with the reasoning of Eckert stating that local and district authorities try to avoid risk and just follow the orders from above. The former state and collective farm manager had the chance to shift into the role as manager of the farm organisation. This would allow him to keep his/her position within the social hierarchy without sharing the same risk of being responsible for the unproductivity of the farm. Hence, he/she does not have the same obligation to the top. Eckert argues that the state and collective farm manager can also lose his/her legitimacy to the employees of the former farm. In her example from the fieldnotes 'the chairman of the kolkhoz undermined his legitimacy and authority by unjustly distributing land' (Eckert 2000, p. 38). Again this is one example, in none of the water user associations visited the farm organisation managers had a reduced legitimacy or authority. Even though, land had been unjustly distributed. In the situation of Karakalpakstan with extreme water scarcity and low productivity farms, state and collective farm managers could neither guarantee jobs or livelihood security. On these state and collective farms the farmers did not consider the state and collective farm managers as an able leaders, (interview with Bektemirov) consequently their social status declined because of the overall insecurity on the state and collective farms. Therefore the shift from state and collective farms to independent farms eased their responsibility and made the farmers responsible for themselves.

3.11 Changing names without changing patterns of influence

The change from state and collective farm manager to farm organisation manager as well as the shift from head of the hydrological unit to water user association manager seems to be not an actual but rather an intangible change. It seems that under the current structure and the remaining habits of top-down acceptability of orders both jobs keep the same or similar amount of social status within the community. Furthermore, the shift reduces the risk situation determined through the top-down structure and offer security through reducing the responsibility for the farm organisation or water user association manager and dividing the responsibility amongst the newly independent farmers.

The diagram shows that the farm organisation manager still has influence over the supply of goods necessary for the agricultural production and is in a linking position between the independent farmers and other institutions.

It seems that this feeling of security was dominant at the water user associations visited during the first fieldtrip in 2000 and during the visits through Karakalpakstan. During these visits the position of farm organisation manager and water user association manager were not questioned. However, it seems that the acceptance of the position of the farm organisation manager was seen as more secure then the position of the head of the water user associations. In the Beruni water user association in Karakalpakstan, the water user association manager was aware of the fact that the community elected him and that the community would have to decide on his re-election. However, whether the election was democratic cannot be determined, it seems that there where no other candidates applying for the position. Furthermore, if the farm organisation management would have 'nominated' him, then it would have had the same effect as an appointment. According to the water user association manager, he was aware that he exchanged the top-down insecurity to a bottom-up insecurity and that he became a service provider to the users. According to the regularities of water user associations, the actions of a water user association manager should be transparent for the community of farmers and he is accountable to them, hence there would be a bottom-up insecurity. However, neither transparency nor accountability were observable at the visited water user associations. Firstly, the farmers had a lack of knowledge about the duties and rights of water user association members and their representatives. Secondly, the head of the water user association had a top down approach to deliver water, he decided, who should receive water, rather than the community of farmers. Hence, he continued with the habits learned during the time of the state and collective farm and did not try to change water distribution. Furthermore, it is questionable whether the former state and collective farm managers who are now farmers as well, can utilise their status within the community to receive more water. The shift from top-down to bottom up insecurity does not seem to be a real change, because the power stakeholders changed their position as well, the shift did not change their influence and their potential to manipulate.

During the interviews with farm organisation managers, it was not evident that their image of themselves had changed from a manager, who directs his own business, to a service provider who is employed by the community. Hence, the farm organisation managers did not see themselves as an employee of the farm organisations, but continued to react as a state or collective farm manager. This might have been to do with the fact that he was nominated for the farm organisation position from the district government. Hence, the top-down insecurity in terms of job and social status stayed the same. However because he established a farm, he had additional security.

Figure 4.3 Power and Influence of the stakeholders during the second land reform

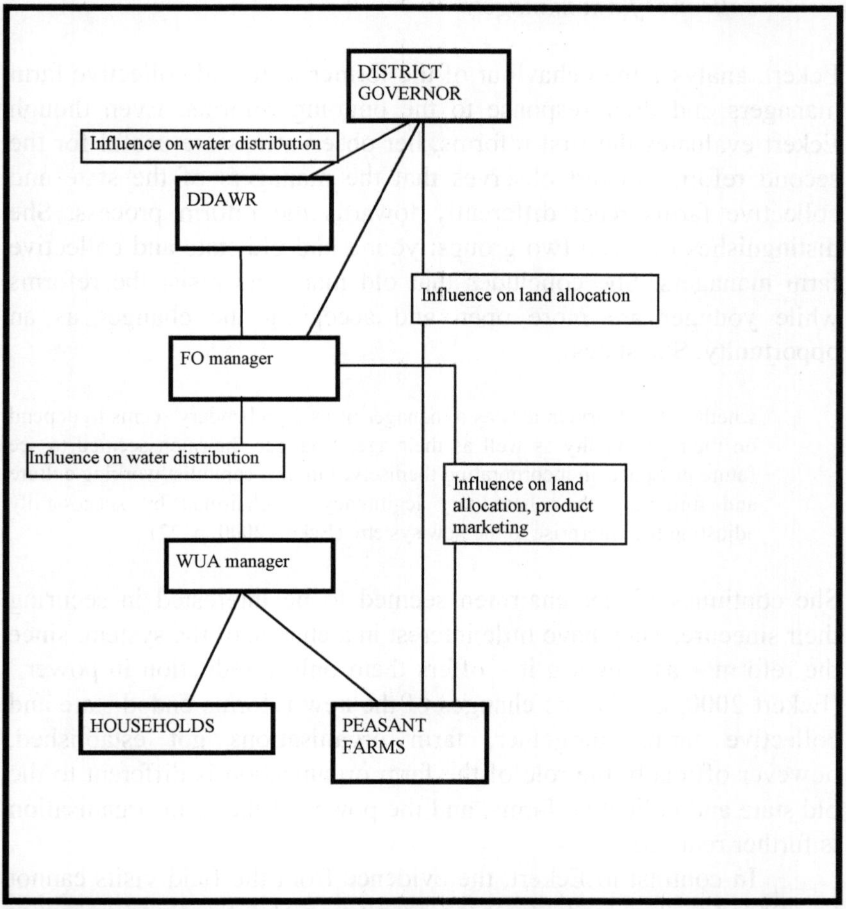

DDAWR: District Department for Agriculture and Water Resources
FO: Farm Organisation

3.12 *Anticipated new roles and responsibility of the state and collective farm managers*

Eckert, analyses the behaviour of the former state and collective farm managers and their response to the ongoing reforms. Even though Eckert evaluates the first reforms, her observations are useful for the second reform. Eckert observes that the managers of the state and collective farms react differently towards the reform process. She distinguishes between two groups: young and old state and collective farm managers. She concludes that old managers resist the reforms while younger are more open and accepting the changes as an opportunity. She states

> whether the chairman acts as a manager or as a prebendary seems to depend on their personality as well as their age. Younger chairmen seemed to see future prospects in incorporating themselves in new capitalist working culture and furthering their hereditary legitimacy as chairman by successfully adjusting the enterprise to the new system. (Eckert 2000, p. 22)

She continues 'Older chairmen seemed to be interested in securing their sinecure. They have little interest in a change of the system, since the reform – as they see it – offers them only a reduction in power.' (Eckert 2000, p. 22) The changes of the new reforms ended state and collective farms altogether, farm organisations got established, however officially the role of the farm organisation is different to the old state and collective farms, and the power of the farm organisation is further reduced.

In contrast to Eckert, the evidence from the field visits cannot be utilised to draw a determining conclusion about age and openness for changes. On the other hand it has to be questioned how her interviews were conducted, and how openness and age is defined. In addition, it is to assume that 'openness' and 'adaptiveness' to changes, might depend on the security or insecurity of the current overall situation. The evidence of the field visits indicate that at the provincial level change was perceived as positive. This was due to the fact that the whole system of water allocation in the Amu Darya basin was perceived as insecure. In different provinces of Uzbekistan the state and collective farms vary in terms of productivity, however in downstream regions especially in Kararalpakstan, because of the water scarcity the overall security of the state and collective farm managers is low. Their insecurity is based on the problems to fulfil the agricultural production plan and secondly to provide labour for the state and collective farm employees. Because the situation is insecure, changes, which will offer some stability, such as establishing a private farm, are likely to be perceived as positive and desirable.

3.13 Gate keepers of resources

3.13.1 Information

As Eckert mentions, farmers recognised their lack of knowledge about farming and management. She states that the farmers expressed the need for training to strengthen their ability to deal with the independent management of a farm. During interviews in Urgench it was stated that farmers contacted the head of the former state or collective farm for advice on current problems. In addition once a month was a meeting, in which farmers could ask more general questions. Hence, the farm organisation manager as well as the water user association manager had to be contacted for information relevant to farming and water use. The farmers were not able to use other sources for information, hence the dependency of the farmers on the 'experts' increased further. Expert 'knowledge' and 'information' were power tools to keep the original hierarchy in place. In addition, the distribution of knowledge could be interpreted positively from the new independent farmers, consequently the social status of the position of the farm organisation and water user association manager increased.

However as already mentioned in the first part of the chapter, the land distribution altered the local level and added new farmers with connections to different levels. This would imply, that these new farmers might have other sources of information and might be more critical towards the old local hierarchy. The two former state and collective farm managers in the Urgench water user association had more possibilities then the other farmers. Both were selling rice from their land not on the local market but in different provinces, where they could make more profit. (At the time it was prohibited for Khorezm farmers to sell their agricultural goods outside Khorezm.) According to the Farm organisation manager in Urgench, farmers had the possibility to buy additional private pumps for their land. However, only the farmers with connections and money could utilise this option, other farmers were dependent on the normal water distribution through the water user association. Hence, connections and money increased their bargaining power, and therefore the possibility to bend or to break rules, or to get special treatment. Bektimirov confirmed, that the former old stakeholders can get special treatment, either because of the networks or because their financial resources. (informal interview with Bektimirov, June 2001)

The rules of water user associations determine that information should be shared and that members should be involved in the decision making process. However, during the field visits in Karakalpakstan, it seemed that neither the farm organisation managers nor the water user association managers consulted with farmers about decisions or processes. For example, in the Beruni water user association the water user association manager decided randomly which of the farms should receive water. Farmers were not involved in the process, and had officially no influence on the water delivery to their farms. Hence, there was no bottom-up flow of information and therefore no participation in the decision making of the water user association.

3.13.2 Water distribution within the water user association

According to the rules within a water user association the water should be distributed between the members on an equal and fair basis. Hence, all members should share the available resources. This would imply the construction of an irrigation plan, which would determine who, when and how much water should be distributed to the farmer.

The water distribution varied among the different water user association. In the water user association in the Syr Dar'ya province farmers worked together in terms of sharing water. They distributed the water according to a time schedule of delivery. Furthermore, farmers traded their water rights. If the farmers did not need the water on the day, which was allocated to them, they exchanged the water allocation day with other farmers of the water user association. This democratic and participatory way of water sharing differed between Karakalpakstan and Khorezm. In Khorezm the head of the Urgench water user associations allocated water to the farmers according to a time-table. The users of one on-farm irrigation canal measured their water intake through pumps. The on-farm irrigation canal was filled for all the users along the canal. This form of water allocation differed from the experience in the southern districts of Karakalpakstan. Here the manager of the water user associations decided every day which farm would receive water. The water user association manager explained, that he inspected all the fields and decided ad hoc which field needed water immediately and which field could wait to receive water. While the decisions on water allocation in the water user association in Urgench was more transparent, allocation in the Southern districts of Karakalpakstan seemed to be arbitrary. It seemed that it would have been easy for power stakeholders such as the former state and collective farm manager and new independent farmers with connections to manipulate the situation and to receive more water. Hence, it appeared that the availability of water influenced the decision making, and influenced the coalitions on the local level. The building of 'elite' coalitions would have worsened the situation for the new independent farmers without connections. On the other hand, the example of Urgench shows that coalitions could be established to build blocks for more equal and transparent water distribution. However, it seems that it became a question of benefits.

In the Kegeili water user association the former hydro-technician was the head of the water user association. The Keigili water user association experienced severe water shortages. Because of the drought the level of the main canal was so low, that no water could flow into the secondary canal. Hence all the independent farmers along the secondary canal had no water. From 88 farms only 37 planted crops and out of these 15 were in better condition. The farm with better conditions had land, which bordered at the main canal. Special pumps were used to pump water from the main canal onto their land. However, for the remaining farmers no effort was undertaken by the water user association to install pumps at the main gate or to divert water from the farms with access to water to the farms with no access. Arguably, the farmers did not have enough financial resources to buy pumps or to divert the water. In addition, it is questionable whether the farmers had enough social capital to utilise the water user association to the benefits of all the farmers. Because, the head of the farm organisation had access to water, it could be argued that the old hierarchy was still in place and obstructed social capital building to take action. A functioning water user association would have affected the farm organisation manager's access to water, hence it was in his interest that the water user association was not functioning. In this case, it could be questioned whether the new independent farmers with connections to different levels would be positive or negative. It seems to depend whether they received land with access or no access to water. Hence, it would be possible that the building of coalitions between farmers might have weakened the total benefit for the whole community.

3.14 *Habits and risk management*

During the first land reform it was reasoned by the administrative body that because of the 'socialist mentality' farmers did not want to establish their own private farms. As reasoned above it was argued that farmers did not want to take over responsibility and act independently. On the other hand, Eckert reasoned that former farm employees wanted to gain more experience, before becoming independent. The second reform indicates that nothing had changed. The main beneficiaries of the reform were the 'administrative bodies' who applied again, and utilised their networks to receive land. Even though, former state and collective farm employees received land, it seemed that the manipulation of the power stake holders continued. Because the state and collective farm disappeared the local level elites had to secure their share of the land as well, hence less land was available for the ordinary former farm employee without connections. In addition, the figures of the visited water user associations indicate that the reform reduced the available land for household plots. Hence, the reforms revoked the previous reforms, which increased food security of household, by enlarging the agricultural land of households.

Again the land reform and the introduction of farm organisations and water user associations indicates that the concept of the 'socialist mentality' is too simple, to provide a reasonable explanation of the changes. It seems that different 'mentalities' coincided, however these 'mentalities' varied with different actors and with different situations. Hence, there is no clear cut definition of the 'mentality' of a person, which would make future actions predictable. It is only possible to determine after the event which 'mentality' was dominant or more relevant for the actor. Hence, the findings would confirm that there are habits, however it seems that some habits are more evident and influential than others.

It seems that a common feature of the local power actors during the land reform was to avoid risk and insecurity, and to manipulate risk situations through the habits and the solutions, which are best known to these actors. According to the theoretical approaches on institutional change and risk management outlined in the theory chapter this is a normal behaviour of actors and institutions. Hence, in the Uzbek case, networking seems to be an acceptable feature to secure the position and livelihood. While it was argued that the 'socialist mentality' implied not only a lack of independence and decision making, but also acceptance of the inability to change the top-down structure and the acceptance of the legitimacy of the structure is common, the evidence of the land reforms and water user associations seem to suggest a more dynamic approach. The evidence suggests that actors are able to manipulate risk situations and to choose certain options, which suit their purposes.

14 Conclusion

In the chapter the top-down institutional changes have been analysed at the local level. The evidence gathered in the study suggests that an organisational change, in this case a change of the constitutional rule affecting land tenure and access to water, in which an organisation is supposed to change from a top-down structure to a bottom-up structure can be obstructed through the actors implementing these changes. The case study shows that actors in organisations are not neutral to change nor do they just follow orders, but they have their own interests.

The implication is that organisational changes bring instability to the status quo of the actors within the organisation and actors involved in the changes. The evidence suggests, that actors use their influence to stabilise their position and to secure their status. However, just focusing on the organisational changes does not give a full explanation of the situation, one has to understand the context in which the organisation operates, and also the individual or stakeholder in the organisation, his/her real and also perceived context and real and perceived securities and risks. The case study showed that because of the difference of upstream and downstream water scarcity during the drought period of 2000 and 2001, the actors in the regions might have responded to the changes differently. Downstream (northern districts in Karakalpakstan) the risk situation was much higher, than the risk situation upstream (southern districts in Karakalpakstan) or even further upstream in Khorezm. Hence, even though the organisational change was the same, the context of the change could be very different. This circumstance could also explain the different behaviour of the actors allocating water, or being more responsive to the changes.

The changes of the organisation affected not only one administrative level, but also different levels, hence different levels interacted. While in the existing literature on land reforms in Uzbekistan it was indicated that higher levels had influence on land distribution, the literature did not give any indication how this might change the power structures or power alliances on the local level. Nor did it throw light on how the affect this might have for collective action and therefore sustainability of common pool resource organisations. In addition, the literature did not show how the context influences decision making and how it might balance power inequalities on different levels. The next chapter will indicate how actors on the provincial level influenced the changes on the local level.

Chapter five: The provincial level

1 Introduction

The chapter starts off with a brief introduction to the major changes in water management since independence and the changes, which are still anticipated from the leading water management organisations in the Aral Sea Basin. The changing political environment influenced not only the structure of the organisation, but also their tasks. The political changes affected the ability of an organisation to fulfil its task of water distribution and operation and maintenance of the irrigation and drainage system. Even though the chapter focuses in particular on the top-down induced changes of the state and collective farms into water user associations, the chapter will show that the wider transitional frame of the political and water managerial environment heavily influenced the implementation of the changes at the local level.

Analysis at the provincial level is important, because even though it is one of many duties of the province managers of water distribution to introduce water user associations within the province, in the case of Karakalpakstan and Khorezm the introduction differed significantly from each other. It will be argued that the wider water management changes are responsible for the approaches taken to introduce water user associations. Hence, the chapter highlights the different issues that motivated the province managers of water distribution and their reactions to political circumstances. A particular interest will be taken in the top-down introduction of water user associations in their respective provinces. It will be argued that the uncertainty of their own position at a different institutional level influenced their actions. Hence, the chapter highlights how perceived uncertainty or risk situations triggered reactions which are complex and dynamic.

As argued in the theory examined in chapter two, organisations operate at different levels, and changes of the organisational structure or the power relations within the organisation affect agents at different levels differently. This chapter pushes forward and tests the reasoning in the theoretical chapter in relation to different organisational levels and how stakeholders on one level influence changes on lower levels. It will show how the province managers of water distribution will influence the establishment of local level water user associations. In the literature it was reasoned that institutional change is dependent on the benefits of the stakeholders implementing the changes. It was suggested that these stakeholders are positive towards changes if and only if they are affected by these changes positively. Here it will be shown that changes on one level are interpreted on a second level, and cause reactions at the second level that influence the implementation of the changes of the first level. It will be shown in this chapter how the province managers of water distribution interpreted the establishment of water user associations as an opportunity for them to stabilise their own positions by interfering in the process of implementing water user associations.

2 Organisational background

Uzbekistan has a Ministry of Agriculture and Water Resources. The presence of a Ministry already indicates the significance of water resources for the country. As stated in chapter three nearly 92 per cent of water resources are used in the agricultural sector for irrigation. The Ministry on the national level has representative offices on the provincial and district levels: in Khorezm the department for Agriculture and Water Resources and in the autonomous republic of Karakalpakstan the Ministry of Agriculture and Water Resources are responsible for water distribution. Even though water and irrigation is still a high priority, it seems, that during the period of the Soviet

Union, water and irrigation had a higher priority. During that time irrigation and water resources had its own Ministry. After independence and during the transition the power of the old water institutions declined, and irrigation and water resources became subordinated to agriculture.

Prior to independence, the decisions on water resources and development were made in Moscow and not in the Uzbek SSR. The decision-making on water was top down which reflected the interest of the USSR and not the interests of the local, district, province or Soviet republic level. This implies at the time of the Soviet Union, that the Soviet rivers were managed according to basin boundaries and not with reference to smaller administrative boundaries. The hierarchy of the institutions making decisions on water management differs. According to this perspective water could be either a technical-managerial or a political issue. While Renger starts with the highest political institution the Central Committee of the Communist Party of the USSR, the SIC ICWC report starts with the Ministry of Land Reclamation and water resources of the USSR. This shows that depending on the perspective of the involved party the hierarchy differs. In this analysis, the perspective is that water is political, therefore Renger's structure is emphasised, which includes the Central Committee of the Communist Party. The party in Moscow decided on the political strategy. Their strategy had consequences for the decisions at all levels on water and land. The chosen strategy was not only political but was influenced by the 'hydraulic mission' which inspired the policy in Moscow from the 1930s to 1989.

Based on the strong hierarchical system, the following institutions tried to implement the decisions taken by the Central Committee. According to Renger, the third institution in the hierarchy was the Ministry of Land Reclamation and Water Resources of the USSR. The ministry had its subordinate organisations at the republic, provincial, and district levels. At the bottom of the hierarchy were the water users at the farm level (for details see figure 5.1. Institutional Water Management Structure in the Soviet Period).

Figure 5.1. Organisational Water Management Structure in the Former Soviet Union

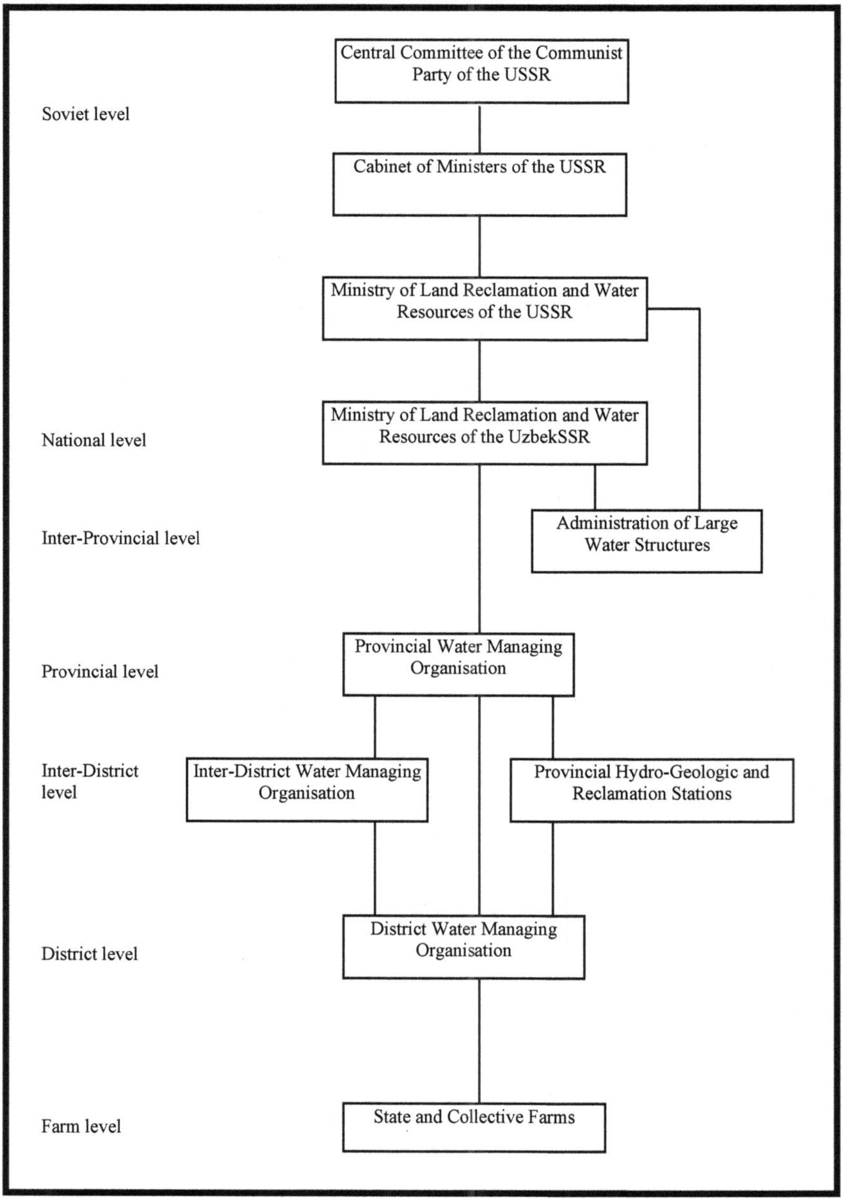

After independence, the pre-independence water allocations to the republics continued. The rivers of the Amu Darya and Syr Darya continued to be managed on a basin scale. The River Basin Organisations (BVOs), which were created in the 1980s continued being responsible for water allocation and the monitoring of the water management structures. Even though they are supposed to have the same power as before, de facto their status and possibility to operate has been reduced. Within the Uzbek republic, water is managed by a strong institution with a top down approach. In Uzbekistan as elsewhere the water management arrangements differ on the basis of whether they are viewed according to a political or a technical perspective. According to Renger on the top of the hierarchy is the President of Uzbekistan followed by the Cabinet of Ministers and then the Ministry of Land Reclamation and Water Resources. This Ministry gives orders to the Provincial Department for Land Reclamation and Water Resources, which is superior to its District Department and the farms at the local level.

In autumn 1997 the Ministry of Land Reclamation and Water Resources merged with the Ministry of Agriculture to create the new Ministry of Agriculture and Water Resources. The structural change at the national level was continued at the provincial and district levels. Hence, the merger affected all administrational levels, and the changes were implemented from top to bottom. Within the new institution, however, the separation between agriculture and water resources structure still remains. The Ministry and the departments are divided into two separate units. The Minister is from the agricultural unit and his deputy is from the water management unit. The structure indicates that water has still a high priority, but is now subordinate to agriculture and is not anymore independent. The officials responsible for water allocation are the deputy minister or the deputy head of department for Agriculture and Water in the province.

SIC ICWC confirms that the merger affected the power and influence of the water institution negatively, within the institution water became subordinated to agriculture.

> At provincial level after merging into one provincial administration of agriculture and water resources – Oblselvodkhoz – level and means of water related organisations decreased sharply, they lost their independence with arisen subordinated state. (SIC ICWC 1999, p. 20)

According to SIC ICWC one consequence of the merger has been that the influence of the water institutions declined with respect to the organisation of works at district and provincial levels. Examples of the shift to an Agriculture and Water ministry influenced the provincial water infrastructure as well as the administrative-territorial organisations.

SIC ICWC listed the following functions for the Oblselvodkhoz: '(1) drawing up of annual water balances for province and water allocation between systems with establishment of limits for a year, vegetation period and months; (2) control of water supply to water users according to established limits; (3) organisation of inter-farm irrigation and drainage network and hydrotechnical structures operation as well as maintenance of the system according to the contracts with water users; (4) keeping annual accounting of the irrigated lands reclamation state, the quantity and quality of return waters; (5) equipping state water systems with water meters and accounting devices as well as their metrological provision; (6) taking part in selection of priorities in new water structures planning; (7) development and implementation of the measures on water structures technical state improvement'. (SIC ICWC 1999, p. 21)

Figure: 5.2 Water Management in Uzbekistan

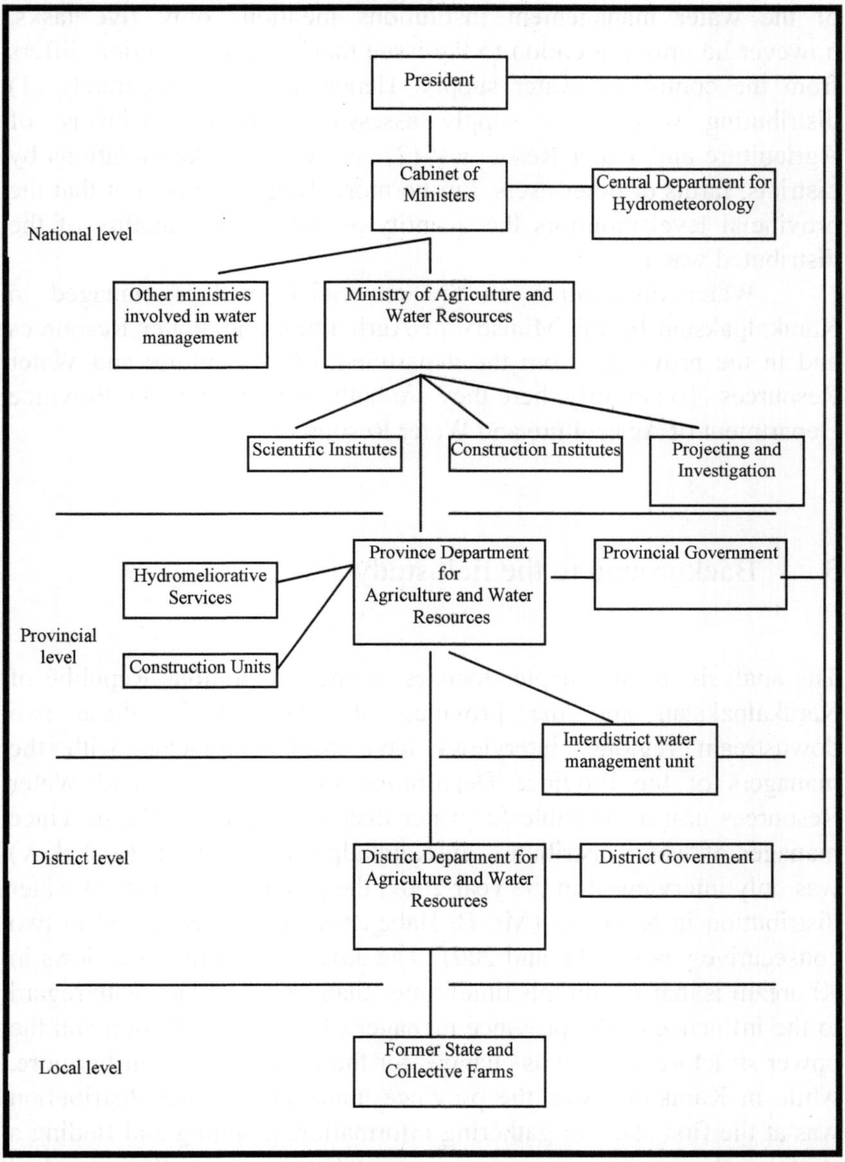

While SIC ICWC points out these seven tasks, Renger in his analysis of the water management institutions mentions only five tasks, however he draws attention to the issue that water distribution differs from the control of water supply. Hence, he states separately (1) distributing water after supply assessment by the Ministry of Agriculture and Water Resources; (2) control of intake violations by districts, farms or other users. Furthermore, Renger points out that the provincial level monitors the quantity as well as the quality of the distributed water.

Water distribution on the provincial level is managed in Karakalpakstan by the Ministry of Agriculture and Water Resources and in the provinces from the department of Agriculture and Water Resources. To simplify, here they are both referred to as the Province Department of Agriculture and Water Resources.

3 Background to the fieldstudy

The analysis in this chapter focuses on the autonomous Republic of Karakalpakstan and the Province of Khorezm. In these two downstream regions, interviews have been conducted with the managers of the Province Department for Agriculture and Water Resources unit responsible for water distribution. While the province manager of water distribution in Karakalpakstan (Mr. R. Koshekov) was only interviewed in the year 2001, the province manager of water distribution in Khorezm (Mr. B. Babejanov) was interviewed in two consecutive years, 2000 and 2001. The advantage of the interviews in Khorezm is that within this timeframe, changes took place with regard to the influence of the province manager of water distribution and the power structure, as well as changes for the lower levels. Furthermore, while in Karakalpakstan the province manager of water distribution was at the first stage of gathering information, planning and finding a strategy, the province manager of water distribution in Khorezm had its strategy already established and used its influence to make its plans a reality. Both province managers of water distribution had no communication about their strategies and plans, and both were interviewed independently.

At the time of the fieldwork the changes in these two regions in terms of privatisation and establishing of water user associations were more rapid than in any other regions of Uzbekistan.[23] Changes in water management at the state and collective farm levels create risk situations for the organisations at the local level, and awareness of such risk could trickle up, affecting the organisations at higher levels, such as the district and provincial levels. As a result the positions of the current stakeholders at these levels become uncertain, and their status might continue to become less influential. The impact of the international community and the regional water organisations, such as the Scientific Information Centre (SIC) could also change the structure, power and influence of the established River Basin Organisations, and of the province and district water organisations.[24] A similar development started in May 2001 in Kyrgyzstan. (There new organisations were created in a top-down approach to support water user associations with technical advice at different levels).

The current water institutions are under external and internal pressure. Since independence, the pressure from the international agencies to change the water management organisations has been high. Within Central Asia, it is questionable whether a system of levels such as private farmer, collective farm, district, province, state and basin will continue in the future, or whether some levels will disappear completely. As a result of outside pressure and internal

---

23 However, with the events from September 11th this might change, for example USAID wants to intensify the work on WUAs, the World Bank is financing a project in the Karshi steppe, which will include social changes in terms of establishing WUAs.

24 Berkoff argues that in the past 'the BVOs have weakened partly as a consequence of outside influence. The World Bank etc. may talk of strengthening them but their actions have had the opposite effect. (Personal communication D.J.W. Berkoff 27.06.2003)

uncertainties the established structure is unstable from the top and from the bottom. The process of transformation creates uncertainty for the stakeholders at different levels. In addition, the integration of Afghanistan into the institutional framework of water management in the Amu Darya Basin might have further implications. Not only might there be a reduction in available water, but in addition on issues of water allocation and management. For example a framework could develop distinguishing between upstream, midstream, and down-stream instead of nation states. (Compare to final report of workshop on Afghanistan June 2002).

According to the theoretical framework, introduced in chapter two, the stakeholders at different levels have either the possibility to manipulate or to enhance top-down and bottom-up induced changes. It was reasoned that in both cases the stakeholders will be interested to create stability for their positions without losing socially or economically. It will be shown in the discussion in the following sections how the province manager for water distribution tried to create a stability during the time of uncertainty and institutional change.

4 The current role of the province manager of water distribution

The duties of the Province Department of Agriculture and Water Resources have been outlined above. In the analysis in this section the main focus will be on his most important function which is water allocation. The province manager of water distribution allocates water to the different districts within the province. The first priority in water allocation is given to municipalities. Agriculture comes second. The province manager of water distribution allocates water to the districts. However, not all districts receive the same amount of water. The amount of allocated water varies according to the water needs of the district, which are mainly determined by the crop produced in the district. J. Sulaimenov, the district manager of water distributionof the Kungrad district, a northern district in Karakalpakstan, stated that water is distributed at the provincial level taking into account the national production plan. Because Uzbekistan focuses on cotton production for export the province manager of water distribution allocates water to the districts in which cotton is mainly produced. Hence, water is not distributed equally amongst the users, but according to the goal to fulfil the production plan. Renger draws attention to the provincial and district governors (hokims), who are appointed by the President and are directly subordinate to him. Renger argues that their future is linked to the fulfilment of the cotton and wheat production norms. He states, 'it is reported that some hokims have directly interfered in water management by-passing the Ministry of Agriculture and Water Resources'. (Renger 1998, p. 17) However, the province manager of water distribution and district manager of water distribution are obliged to fulfil the norms and will try to achieve the anticipated plan.[25]

---

25 Putting the functions of the water organisation aside, agriculture and water in Uzbekistan are linked and decided on the top. Hence, even though the water

Even though it is stated by the SIC ICWC that it is the responsibility of the province manager of water distribution and the district manager of water distribution to maintain and operate the inter-farm canals and their structures, it is questionable whether the Province Department of Agriculture and Water Resources or the District Department of Agriculture and Water Resources have the resources to maintain or operate the canal structure. A field visit in the Kungrad district, showed that a small sluice diverting water to different state and collective farms was not functional because of the silting up of the gates. The district manager of water distribution was aware of the problem and pointed out that the state and collective farms took the initiative to free the sluice to keep it functional. Hence, it seemed that the District Department of Agriculture and Water Resources did not have the resources or the capacity to provide efficient operation and maintenance of the interfarm canals. Therefore efficient and equal distribution of the scarce water resources to the state and collective farms was not possible. The lack of resources and capacity is confirmed by O'Hara. She states, that the

> Central Asian Republics are either unwilling or unable to meet the enormous cost of managing and maintaining it. A marked reduction in the number of people working in the water management sector has further exacerbated the situation, and there is almost a complete lack of necessary equipment for maintenance purposes (O'Hara 2000, p. 367).

The field study confirmed obligations of the province manager of water distribution outlined by Renger and SIC ICWC. The province managers of water distribution in Karakalpakstan and Khorezm were in charge of water allocation and distribution to the districts. Furthermore, they had to control the water intakes of the districts as well, to ensure that the upstream districts did not take more water than allocated. According to the district manager of water distribution in the Kungrad district, the province manager of water distribution had to

---

organisation still has certain functions, water distribution to areas and farms is politically coordinated.

allocate water in 10-day intervals. Renger verifies this for the district, he states that the supply plans for water delivery to the farms 'is broken down in 10-day periods.' (Renger 1998, p. 15)[26]

In Karakalpakstan, the province manager of water distribution introduced a system of monitoring to control upstream and downstream water use of the districts. The monitoring system introduced took into account the different interests of upstream and downstream users. Water inspectors from downstream districts controlled the pumping systems of upstream districts. However, because of the hierarchy of the system, in times of the drought the province manager of water distribution had to exert his position and his authority to enforce the restrictions on upstream users in the province.

At the district level these measures of control were not imposed. Only administrational staff from the district were responsible for the control of sluices, hence according to the province manager of water resources the level of misallocation was higher. Because the districts reinforced the allocation policy of the province, farms producing rice were not supposed to receive water. All the available water was allocated to cotton farms. However, cotton farms on the district level were in competition over the available water resources, hence, the drought increased the informal arrangements on the district level over water allocation and distribution. This became evident in the two visited districts in Karkalpakstan, Beruni and Kirgili district.

The field study confirmed that water distribution in the province and the districts was related to managerial questions as well as political issues. The situation of uncertainty and the fusion of technical and political issues made the position of the province manager of water distribution even more difficult.

---

26 Officially the 'water demand is calculated on the basis of the following indicators: (a) Structure and area of cropped land; (b) Irrigation regime (method, norm and period); (c) Efficiency of irrigation network and irrigation technique.' (WB 1996, p. 69, compare also Tacis 1997, p. 13) The BVOs supply water in response to national demands within a previously determined plan.

In addition to the obligations of the province managers of water distribution mentioned by Renger and SIC ICWC the province managers of water distribution in Karakalpakstan and Khorezm were responsible for the introduction of water user associations in their provinces. In the first encounter with the province manager of water distribution in Karakalpakstan, it seemed that he understood his responsibility towards the water user associations to be limited to knowing about the statistics of water user association establishments in the province. However, at the first meeting with the province manager of water distribution in Khorezm, it was evident that he gave more attention to the details of the establishing water user associations and was aware of their performance.

5 Motivation and strategy to manipulate the risk situation

In Karakalpakstan the province manager of water distribution anticipated that his position as manager was already unstable and would be even more unstable in the future. Even though he had been in his position for 13 years, there had been considerable changes in recent years in terms of his status, and orders from above (informal interview, Mr. R. Koshekov, July 2001). It seems that Mr. Koshekov gained the job during the cotton scandal in the late Soviet period. The district managers of water distribution of the districts visited in Karakalpakstan were young and belonged to a new generation. The old elite from the pre-independence period had been partly dissolved. In informal interviews at two District Department of Agriculture and Water Resources in Karakalpakstan, it became evident that the district manager of water distribution knew the province manager of water distribution from the University and had close ties with him. This confirms Jones Luong's statement concerning elite training in the same institutes and the building of an additional 'layer of camaraderie' (compare chapter three). Furthermore, according to the province manager of water distribution of Karakalpakstan some former state or collective farm managers in the province had been replaced from above. It was feared that this could happen on higher levels as well, in case of problems with the fulfilment of the production plan or because of arbitrary decision making on the top. This confirms similar experiences cited by Eckert. In Karakalpakstan it was anticipated that replacements at higher levels could happen. However, since independence no staff had been replaced at the district or at the provincial levels. It seems that especially during the period of drought the under-performance of the agricultural sector of the downstream regions became more evident. (compare chapter three and for more detail Wegerich 2002c).

In addition to the uncertainty concerning status and position, the province manager of water distribution stated that his authority was restricted. According to him, he was unable to change situations because the structure of the system was too established and confining for bottom-up or top-down changes at his level. However, the structure was flexible enough for the small changes in regard to water control mechanisms upstream, which he anticipated and implemented. Because he perceived the water system too restrictive and non-dynamic, he was looking for ways to reform the water sector from outside the system, hence creating a new structure. He was planning the structure, which he wanted to create, which would be was more flexible. However, his strategy was to gain a similar position within the new structure. Hence, the structure was supposed to become more flexible, but not in terms of his own position. His priority was to achieve a structure, which would establish security for him. It appeared that flexibility implied only flexibility for his security and not flexibility for the whole system. In addition, even though he wanted to create a new system, implying outside changes, it would be questionable whether these changes could be considered as coming from the outside, or whether it would be a replication of the existing structure with minor changes.

In comparison to the province manager of water distribution of Karakalpakstan the province manager of water distribution of Khorezm seemed to be more aware of the larger context of water management in Uzbekistan, and the anticipated changes of water management. He perceived the structure of the water management system as unstable. (informal interview with B. Babejanov, June 2000 & July 2001), This was for him the primary reason to look for other options. He stated that the water distribution system, as it was established at the time of the interview, might be unsafe in the future. The water management system was organised from above. The highest level determined the allocation of water. Furthermore, because Khorezm and Karakalpakstan are downstream regions their available water decreased after independence, consequently the agricultural productivity of the region declined as well (compare Renger 1998; Wegerich, 2002b).

The province manager of water distribution of Karakalpakstan wanted to utilise the governmental decision to establish water user associations on the local level. He anticipated that water user associations had to be managed and co-ordinated at the district level as well as the province level. Hence, he was planning to create an NGO, which is responsible for the water user associations on the province level. However, he was not certain about the procedures of establishing the NGO, and about whether the anticipated structure would be functional and accepted by insiders and outsiders. It seemed that the province manager of water distribution of Karakalpakstan did not have outside information, beyond his own province with respect to water user association establishments or structures beyond the water user association. In case he had information, it seemed that the information was based on the numbers of water user associations in the neighbouring province. However, he was unaware of studies contacted in Uzbekistan or of the problems faced in different provinces.

For the province manager of water distribution in Karakalpakstan a window of opportunity opened when a research team working on water user associations in Uzbekistan consulted him on issues concerning the establishment of water user associations in the republic and the success of these organisations. The research team included members from a British university (SOAS), the Uzbek Academy of Science, and a national NGO. During formal and informal meetings information exchanges took place. The research team informed him about their experience of water user associations in other countries and the experience in other regions in Uzbekistan. The province manager of water distribution got involved in discussions about sustainability issues of water user associations. He started to disclose his ideas about an umbrella organisation to the research team in return he got information about similar organisations in other countries. The province manager of water distribution also proposed to collaborate with the research team on establishing such an umbrella organisation together. Hence, he wanted to employ their experience and prestige, and to utilise their connections to apply for funding outside the governmental structure. He was looking for alliances to decrease anticipated difficulties. The research team as outsiders to the structure, had access to different resources and information that would enable him to keep his current position, and also to build up the structure as an insider from the outside.

6 Anticipated new job and responsibility

Prior to the contact with the research team, the province manager of water distribution in Karakalpakstan planned to create an NGO, which would be responsible for co-ordinating local water user associations and their demand for water. He planned to create a position for himself, which would have been similar to his current position as head of water distribution at the province level. Therefore, issues, with which he had already experience such as conflict prevention and conflict management, were included in the proposed co-ordination. However, because of the influence of the research team and their agenda, he additionally incorporated the idea of creating a support unit, which was still responsible for co-ordination and distribution of water, but had a strong component of training and assistance to the already existing and the newly established water user associations.

The anticipated plan of the province manager of water distribution in Karakalpakstan indicated that the structure influenced his perception of the new job and its responsibilities. Hence, it would confirm the assumption of Douglas that significant changes do not come from within the structure. However, because of the interest and the information from the research team he shifted his plans for the new structure to find new alliances and to create a higher level of security for himself.

While at the first meeting in 2000 with the province manager of water distribution in Khorezm, the province manager of water distribution emphasised his need for information about equal water distribution and control mechanisms to prevent water theft and over utilisation, as well as a focus of technical solution to water management. In 2000 the training of water user association members to make the farms and the organisation more sustainable in terms of management and participation was not on his agenda. However, during the second meeting in 2001 his emphasis had shifted. The province manager of water distribution had organised training sessions for water user associations. Furthermore, he had established an organisation, which was planned from the province manager of water distribution in Karakalpakstan. It was registered as an NGO and functioned as a support unit for the existing and newly establishing water user associations. It seemed that the training sessions of members of the water user associations in Khorezm were developed from this organisation. (compare chapter four) However, the boundaries between the Province Department of Agriculture and Water Resources and the NGO were not clear-cut. It is not clear whether the province manager of water distribution helped to establish and to train the water user associations as a governmental official or whether he acted as the head of the NGO, which was located in the same building as the Province Department of Agriculture and Water

Resources. Because of his double function the roles of water distribution department and of the NGO were interwoven and the overlap of one with the other was blurred. Hence, it is not possible to state whether the NGO had already taken over some of the functions of the Province Department of Agriculture and Water Resources or whether this was anticipated in the long run.

It seemed that the NGO functioned as a support organisation for the establishing water user associations in Khorezm. It combined the two advantages, which were mentioned above, more flexibility and access to outside information beyond the structure. (see below) These NGO was able to bring benefits in establishing water user associations. The Khorezm water user associations received training from outside the local level of administration, and received information, which was not even accessible to the farm organisation managers. The NGO staff in effect undermined the positions of the farm organisation managers and made the water user association members more equal. In addition, the NGO enhanced the mobility of the province manager of water distribution in Khorezm as will be shown below.

7 Structures of planned support NGOs in Khorezm & Karakalpakstan

While in Karakalpakstan the province manager of water distribution mentioned explicitly how he expected the support unit to be organised, in the case of the province manager of water distribution of Khorezm the structure has to be deduced from his behaviour and from the non-involvement of the district managers of water distribution at any stage of the field visits.

The structure preferred by the province manager of water distribution of Karakalpakstan was almost identical with the current hierarchy of the system. He distinguished among three levels: the province, the district, and the local levels. This structure is similar to water user association support units in Kyrgyzstan. In 2001 an ADB project implemented the first units at the national and at the provincial level. It was expected that supprot units would be added down to the district level. The structure preferred by the province manager of water distribution was confirmed during meetings at the local level. The province manager of water distribution arranged that the meetings were attended by the district manager of water distribution, in order to get them further involved in water user association establishments and to support the ties between the provincial and the district levels. In addition, the province manager of water distribution of Karakalpakstan organised that a district manager of water distribution from a southern district of Karakalpakstan to accompany him and the research team to a meeting on water user associations with the province manager of water distribution of Khorezm in the neighbouring province. Hence, he took some trouble to involve the district level. However, because of the close ties of the province manager of water distribution with the district managers of water distribution it is difficult to judge whether he expected the new structure to be a reproduction of the old structure, or whether he had other reasons, or did not really think about it.

Figure 5.3 Anticipated Structure of the WUA Support Unit NGO in Karakalpakstan

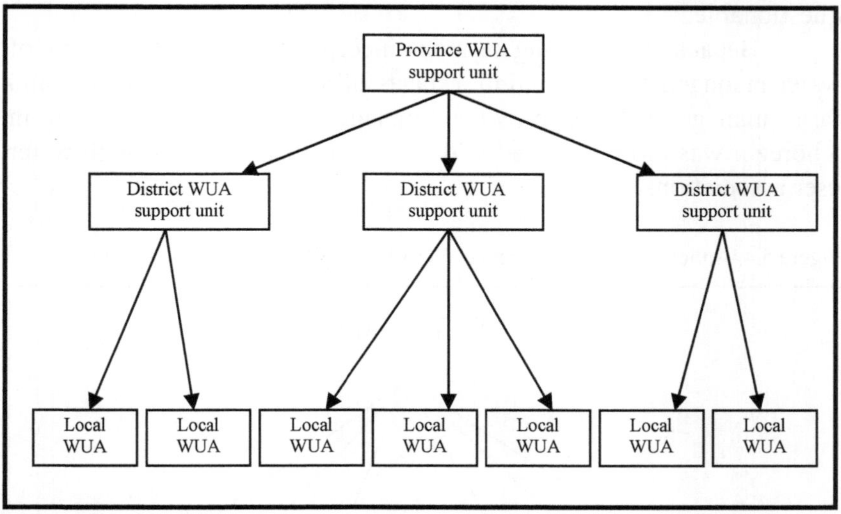

WUA: Water User Association

In Khorezm the province manager of water distribution did not explain the structure of the support unit. However during the visits to the water user associations in Khiva and Urgench the district managers of water distribution of the respective districts were not present. Hence, it seemed that the structure incorporated only one unit on the provincial level that was directly linked to the local level. Because the number of the water user associations was still small, it seemed that it was unnecessary to involve the district levels. However, because the province managers of water distribution of Khorezm and Karakalpakstan expected that the number of water user association in their provinces would continue to rise in the near future, the early participation of the districts would have had advantages for the water user associations. In addition, because in Khorezm it was stated that the water user associations should receive training during the non-irrigation season, it would be necessary to involve experts from the district level rather than experts from the provincial level. The non-involvement of district experts made the success and the equal spread of training sessions for all water user associations in the province questionable.

Because the training sessions incorporated not only aspects of water management, but mainly aspects of agricultural production and farm management the province manager of water distribution in Khorezm was more involved with farm organisations than with water user associations.

Figure 5.4. Anticipated Structure of the WUA Support Unit NGO in Khorezm

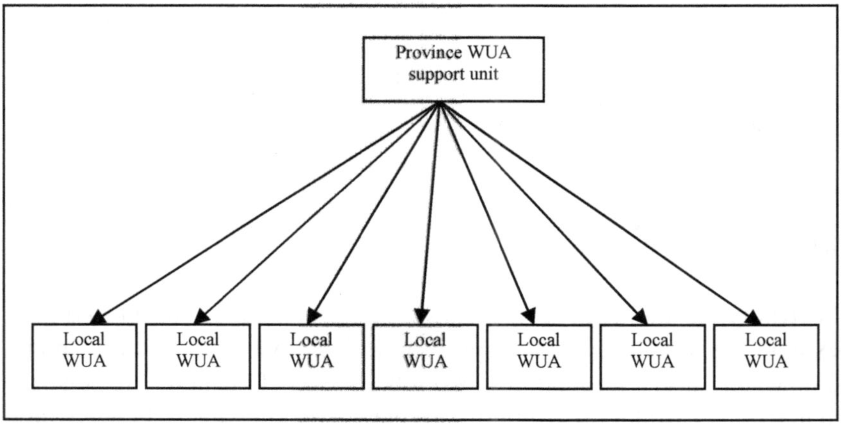

WUA: Water User Association

8 Gate keepers of resources

In the theoretical chapter a gatekeeper of resources was defined as an individual who controls: networks, the information flow as well as the upward, and outward mobility of individuals. Information can be internal or external. Hence, the gate keeper controls what kind of external information reaches the inside and what kind of internal information reaches outside. Hence he/she can manipulate perceptions of internal and external events, actions and structures.

The analysis will analyse whether the province manager of water distribution in Khorezm and Karakalpakstan are gate keepers and to what extend they are using their power position to influence action arenas of individuals. There will be an attempt to determine whether they manipulate resources and if so to what extent.

8.1 Internal information

In Khorezm and in Karakalpakstan the province managers of water distribution were to be responsible for the establishment of water user associations. Because the province managers of water distribution were responsible they had the order from the national level to be in charge of the local level changes. However, it was the responsibility of the province managers of water distribution how to implement the changes. While in Karakalpakstan the province manager of water distribution passed on the responsibility for water user association establishments to the district and the local level, in Khorezm the province manager of water distribution chose to retain responsibility at the provincial level. The district and not the provincial level would have been more appropriate to handle the water affairs at the former state and collective farm level. Hence, it was a personal decision from the province managers of water distribution whether to keep the responsibility or to pass it down, this became evident because of the different forms of implementation, which took different forms in Karakalpakstan and Khorezm. As mentioned above, the province manager of water distribution of Karakalpakstan invited district managers of water distribution to accompany him and the research team to visit different water user associations within Karakalpakstan or even to Khorezm. The district managers of water distribution were present when members and heads of water user associations were interviewed and commented on the organisations and on the levels of participation of the members. Furthermore, the province manager of water distribution in Karakalpakstan shared his views on the established water user associations with the district managers of water distribution. The province manager of water distribution treated the district manager of water distribution equally and tried to involve all of them in the establishment of water user associations. On the other hand, the province manager of water distribution in Khorezm did not want to share information or involve the district levels. While in Karakalpakstan a discourse took place between the province and the district water managers, in Khorezm there was no participation of different levels and no discourse.

As mentioned above, the province manager of water distribution in Khorezm did not involve the district level in the process of establishing water user associations or to support the newly independent farmers. His contact was direct to the local level, to the former state and collective farm managers, but also to the local farmers. The contact was on a regular basis. The province manager of water distribution was in control of the idea of water user associations and training sessions. He promoted these ideas and gave advice to the state and collective farm managers, but also to individual farmers. who before land reforms had direct ties to the Province Department of Agriculture and Water Resources or were members of staff in the Province Department of Agriculture and Water Resources. In Khorezm the establishment and guidance of water user associations excluded one existing level, the district level. In contrast in Khorezm, the province manager of water distribution in Karakalpakstan seemed to rely on the districts and their involvement in water user associations formation. The province manager of water distribution did not have direct connections to the farmers, but had to rely on the existing structure.

The structure of the support unit in Khorezm had the consequence that the new managers of the farm organisations and water user associations had to seek information and advice from the province manager of water distribution. Hence, it seemed that the local level had to bypass the district level, because the district level was not involved in water user association establishments. The type of information included strategies for sustainability, conflict resolution and about the organisation of water user associations. However, because of the organised training sessions, it seemed that the province manager of water distribution was also the person responsible for trickling down information regarding questions about farm management. In return, he received information about the developments of water user associations in Khorezm, their problems, challenges and possibilities of formation and their day-to-day routines. Because of the exclusion of the district level the province manager of water distribution had the most information over the situation of water user associations in Khorezm.

The province manager of water distribution's behaviour could be interpreted in two ways. First, it is arguable that there were only a small number of water user associations established in the Khorezm province. Because there were only six water user associations, the district level did not have to be involved. Arguably the involvement of the district level could have made the process of establishing water user associations too bureaucratic. On the other hand, according to the policies of the Uzbek government more water user associations will be established in coming years. Therefore, there will be a need to involve lower levels to co-ordinate water user association establishment and the training programmes in the future. It could be argued, that the province manager of water distribution in Khorezm wanted to be in control of the development of water user associations and tried to exclude possible competition. He controlled the resource information on water user associations in Khorezm. While former state and collective farm managers had information about one water user association, he had information over all six water user associations in the province. Hence, he was in control of the information about the development of water user associations. He had the monopoly of bottom-up information. If there had been a district level involvement, then it would have created a competitor with information about more than one water user association. Furthermore, if the district level had been more involved, then it would be questionable whether his presence and guidance would have been needed at the local level. Hence, his information would not have been that detailed.

The evidence suggests that the head of water distribution tried to control the resource 'local information'. Furthermore, with his tactic of excluding one level, he reduced the upward and outward mobility of one level, the district level. Because the heads of the farm organisations or the water user associations had only experience with one water user association, they were not in the position to threaten the power position of the province manager of water distribution. Keeping in mind, that he was uncertain about the future of the water distribution system, he succeeded in gaining an information advantage, which helped him to achieve bargaining power for future institutional changes and for his importance to outsiders.

8.2 External information

It is difficult to demonstrate that the province manager of water distribution had information, which they did not distribute to lower levels. However, it might be possible to show that because they had a growing interest in the development of successful water user associations, they became keen to get more information on water user associations and to act on the information. However, again it would be difficult to prove that they had knowledge, which would be negative for their own position, and that they protected themselves by not distributing information. Hence, the awareness of a possible risk situation and their reaction is difficult to prove.

In the year 2000 during the interview with the province manager of water distribution of Khorezm it became evident that he had no access to information about water user association experience in other countries and regions. However, in 2001, he had this kind of information, as well as information about other water user association projects in Uzbekistan. By the first visit, it seemed that he was familiar with the problems faced by farmers in water user associations, such as equal and efficient distribution of water, free riding and monitoring. However, he was looking for technical solutions, which could be enforced by the water user association manager, and not for institutional solutions, which could be enforced through participation in decision-making and self control. By the second visit, a year later, it seemed that he was attentive to the problems of newly independent farmers in transitional societies, meaning the problems of shifting from the status of state farm employee to independent farmer. The province manager of water distribution of Khorezm understood, that the newly independent farmers were in need of training in farming methods. The organised training sessions reflected the shift in awareness. (see below) On the other hand the province manager of water distribution in Karakalpakstan was unaware of other countries' experience with the water user associations. Consequently, he was unaware of water distribution problems at the local level and accepted to a larger extent the command management from the former system.

In terms of getting access to external information, the province manager of water distribution in Khorezm was more informed about the background of establishing water user associations in Uzbekistan. He was aware that SANIIRI had been contracted by the Uzbek government to study the experience of water user associations in other countries and to establish a blueprint on the formation of water user associations for Uzbekistan. He was also informed about the experience of the first TACIS pilot project in the Syrdarya province. On the other hand, even though the province manager of water distribution in Karakalpakstan was aware of the guidelines, which had been published by the Ministry for Agriculture and Water Resources, he was unaware of the background and how and by whom the information was gathered. Furthermore, he did not have contacts to other national or international organisations, which were or are involved in irrigation management transfer. Hence, his access to information was very limited.

The lack of information of the province manager of water distribution in Karakalpakstan about water user associations experience and the lack of involvement with current issues surrounding the problems of water user associations in Uzbekistan, had the effect that the water user associations in Karakalpakstan were not as functional as the water user associations in Khorezm.[27] Taking the difference between the first and the second visit in Khorezm into account, it seems that the province manager of water distribution had a growing interest in developing water user associations and to reach sustainability. On the other hand, it seemed that the province manager of water distribution in Karakalpakstan had until shortly before the field visit a neutral attitude towards the water user associations. It appeared that he was not involved in gathering new information on water user association developments. It could be argued, that his reduced interest, in accessing information about water user association experiences prevented water user association improvement in Karakalpakstan.

It is difficult to prove, that the province manager of water distribution in Khorezm knew about certain events and outside changes on approaches to water user associations. It is arguable that his idea to scale up water user associations might have resulted from external influences. These influences came either from outside consultants, from the Amu Darya BVO located in Urgench, from the contacts to the national research institute SANIIRI, or from SIC ICWC reports. In other countries water user associations have been successfully scaled up to levels higher than the local level, such as Pakistan and Mexico. SANIIRI would have had information about these experiences. Hence, it might have been possible that he had information of scaling up water user association to different levels, but did not intend to introduce support units on the district level, because it could have undermined his position.

---

27 see chapter on local level changes

8.3 *Outside contacts*

The province manager of water distribution in Khorezm had established good relations with the research institute SANIIRI. Consequently, SANIIRI promoted the water user association experience in Khorezm as functioning organisations. However, a field visit in the year 2000 showed that a water user association in Khiva was not functioning. Initial training courses for farmers were not given, farmers did not have access to information, and did not participate in the water user association. Even though, the water user association performance was low, SANIIRI promoted the Khorezm initiatives and therefore stabilised the position of the province manager of water distribution. Due to the promotion of SANIIRI outside institutions gave more attention to the development in Khorezm. Due to SANIIRI's promotion a consultant from the International Water Management Institute (IWMI) visited Khorezm in the year 2000. The consultant worked on a study on the sustainability of water user associations in Uzbekistan and Kyrgyzstan. (consultant report: Wegerich 2001) This type of visit had the effect that the province manager of water distribution in Khorezm gained access to even more sources of information and could expand his network beyond the national level. Furthermore, even though Karakalpakstan had as many water user associations as Khorezm and was about to overtake Khorezm in the number of water user associations in the year 2002, the reputation of and the knowledge about the water user associations in Karakalpakstan were not as prominent within Uzbekistan and with international organisations dealing with water user associations in Uzbekistan.

The field visits in 2001 to Khorezm indicated that the position of the province manager of water distribution was further manifested. He even promoted the water user association experience in Khorezm in a booklet. Spreading information about the water user association developments helped him to gain more attention from international agencies, interested in promoting projects on water user associations in Uzbekistan. Because he was successful in promoting the case of

Khorezm, the water user associations in Karakalpakstan did not get any attention.

At the time of the second visit the German Centre for Development Research (ZEF) started a project in Khorezm and had contacted the province manager of water distribution to get information about water distribution and water user associations. The USAID Central Asia Natural Resource Management Project (NRMP) also wanted to start working with water user associations. The experience of water user associations in Khorezm was known to USAID, however it did not have knowledge of the experience in the neighbouring republic Karakalpakstan. An USAID expert from NRMP intended to visit Khorezm and to create first contacts for future collaboration on projects regarding the sustainability of water user associations.

The contacts of the province manager of water distribution of Khorezm with outside parties were evident. It seemed that he was perceived at the local level as a person who was able to attract national and international organisations. The field visit showed that not only the changes implemented by him were perceived as positive but already the attention he drew from outside was regarded as significant. Hence, his social status in the community rose with each visit. Furthermore, because he was in control of local level information and had experience with the creation of water user associations in Uzbekistan he gained access to higher administrative levels. These achievements not only stabilised his current position as organiser of water user associations but also increased his upward and outward mobility. In 2002 during discussions with members of the USAID NRMP project it was pointed out that the province manager of water distribution had started to work for USAID in the Khorezm province.

Figure 5.5 Anticipated structure and influence

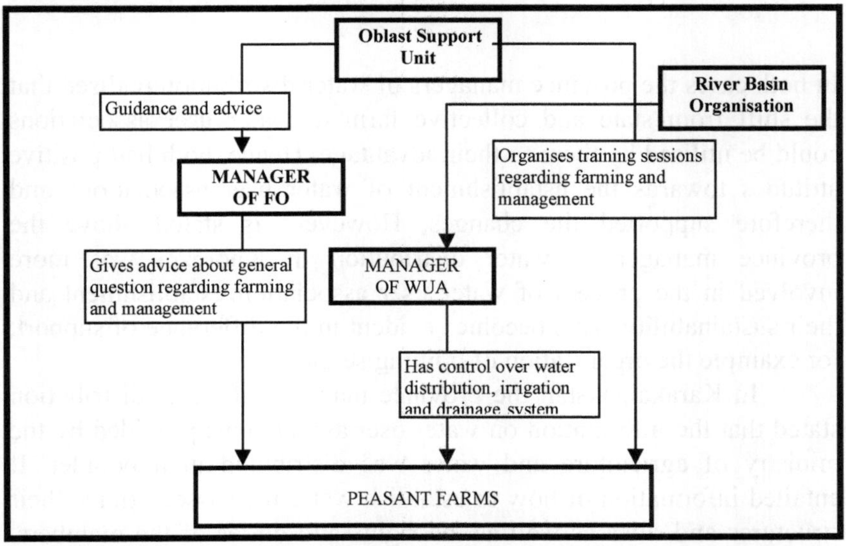

The head of water distribution in Khorezm not only succeeded in excluding the district level at field visits, he also arranged meetings with delegations (in this particular case from Karakalpakstan) away from other people involved with water user associations in Khorezm. At one meeting he tried to establish further contacts with Karakalpakstan as well as finding a combined strategy that would have expanded his influence beyond the provincial level to an international level, incorporating all three downstream provinces of the Amu Darya river basin, Karakalpakstan, Khorezm and Dashoguz (Turkmenistan). In addition, he tried to exchange and to expand his network with other outside agencies. He asserted and made evident his provincial monopoly on contacts. The fact that the province manager of water distribution tried to exclude other water user association stakeholders from Khorezm in the meeting, indicated that he did not want that his power position questioned to them, and could have given them the indication that he was vulnerable and dependent on outside contacts.

In both cases the province managers of water distribution realised that the shift from state and collective farm to water user associations could be utilised by them to their advantage. Hence, both had positive attitudes towards the establishment of water user associations and therefore supported the changes. However, as stated above the province manager of water distribution in Khorezm was more involved in the process of water user association establishment and their sustainability. This becomes evident in the difference of support, for example the organisation of training sessions.

In Karakalpakstan the province manager of water distribution stated that the information on water user associations provided by the ministry of agriculture and water was distributed in a booklet. It entailed information of how to establish water user associations, their structures and rules as well as the rights and duties of the members. However, in Karakalpakstan the booklet was distributed to the state and collective farm managers only, and it was up to the local stakeholders to distribute the information to the independent farmers. For example, it was stated at a water user association in Beruni district in Karakalpakstan that the farmers were informed about their rights and duties as members of a water user association, the same day they established the water user association. In comparison to Karakalpakstan, in Khorezm in 2001 according to the province manager of water distribution every independent farmer received one of these booklets as information. The farmers had the possibility to inform themselves about their roles and duties and were not dependent on the local stakeholders for information. However, it was not stated whether they received the information before, during, or after establishing the water user association. It seems that the farmers received more information to begin with, where the province manager of water distribution was interested in the successful changes towards private farmers and water user associations.

In addition to the lack of information, the water user associations in Karakalpakstan did not receive help in maintaining the water user associations. Until the time of the field visit in 2001 the province manager of water distribution who was responsible for water user associations in Karakalpakstan had not given serious attention to them. Water user associations had been formed and farmers were only partially informed about their duties and rights in the water user associations. Furthermore, even though the first water user associations in Karakalpakstan were established in the year 2000 the farmers had not received any training by mid-2001.

However, because the province manager of water distribution of Karakalpakstan realised the potential for himself if water user associations become sustainable, he was beginning to be more interested to help water user associations to become sustainable. Therefore, he was thinking of organising training sessions for farmers, to increase their abilities as business managers. He planned to train farmers in accounting, taxation, legal aspects and contract formulation. According to him, the newly independent farmers were experts on farming methods and land issues. He argued that farmers knew about the water needs of plants, the fertility of the ground, the problems of salinity, water logging, fertilizer and pesticide use. He argued that farmers do not need any training for farming, that it would be redundant. (Beruni district 29.07.2001)

The field visit in the year 2000 to the water user association in Khiva showed similar results, however the water user association was at the time only eight months old. At the time, the province manager of water distribution in Khorezm made the same arguments as the province manager of water distribution in Karakalpakstan. He stated that farmers did not need training as such, because they lived and worked on a state and collective farm and therefore had the necessary experience as farmers. Hence, for him training was only necessary to increase the business skills of the farmers. However, in an informal interview in the year 2001 he stated that a recent training session included lectures on ground fertility, watering techniques, soil salinity and seeds. He emphasised the importance of these training sessions, arguing that the former state and collective farm employees were only specialised workers, but not full farmers. In addition in the year 2000 there was no evidence that training sessions had taken place, even though it was stated that there had been. In 2001 the province manager of water distribution could provide a time-table of training sessions, which indicated the content and the duration of these sessions. During an interview at the water user association in Urgench district in Khorezm it was stated by the former state and collective farm manager that the farmers wanted to receive training on issues regarding farming and that the training sessions provided reflected the demand of the farmers (Urgench, 29.07.2001).

In Khorezm training was organised from the provincial level. The invited specialists were from the provincial level as well. According to the province manager of water distribution training sessions were held annually during the winter, when farmers had time to attend. The province manager of water distribution also stated that in addition to the annual training sessions special seminars were provided, to show farmers how to leach or how to irrigate. These training sessions took place before water delivery at the beginning to the season.

It seems that the province manager of water distribution in Khorezm realised the importance of farmers' participation and bottom-up developments for the sustainability of water user associations. It appeared that he was trying to accommodate the demands of the farmers where possible. Recognising the fact, that successful water user associations could be utilised by him and could create security for his own position. The second visit showed that a shift in communication took place, while in the year 2000 in Khorezm decisions were made from the top, already one year later the approach had changed and farmers participated more actively.

10 Different influences on the stakeholders

Even though the government ordered the establishment of water user associations on unproductive state and collective farms and the two provinces had a similar amount of established water user associations, the situation and the experience of water user associations was different. The difference was evident in issues such as water distribution and training sessions. The explanation for the difference was first the personal commitment of the province manager of water distribution and secondly his shift in perception. In Khorezm the establishment of water user associations became relevant to the province manager of water distribution before they became relevant to the province manager of water distribution in Karakalpakstan.

The main reason for the lower involvement of the province manager of water distribution in Karakalpakstan, in the process of water user associations formation and sustainability, was that the idea of scaling up water user associations was fairly new. The realisation, that scaling up water user associations could be used to create security at higher levels than the local level, shifted the potential for future involvement and therefore for the future development of water user associations. Commitment to the establishment and training of water user associations was only one part of making these organisations more sustainable.

The commitment of the province managers of water distribution was only one of the reasons, which helped the implementation of the water user associations. The strategy of the province manager of water distribution in Karakalpakstan was still top down, in terms of allocating water to water users, in dealing with conflict situations and in providing training seminars. On the other hand the approach of the province manager of water distribution in Khorezm had already the character of a service facilitation and support unit for water user associations. This approach became even more evident, when it was argued that the members of the water user associations demanded training sessions. The demand and the facilitation of training indicated that the water user associations have already achieved some form of participation.

It is unclear whether the shift from top down to participation is due to the influence of outside information or the realisation of the province manager of water distribution, that only participation and the shift from command management to service provider would stabilise his future security. In any event the shift in his perception of his role seemed to be a significant step in bringing about organisational sustainability. The province manager of water distribution in Karakalpakstan had still not made this shift. It would be interesting to analyse how this shift on the provincial level might trigger other shifts or might increase the demand for more changes on different levels and maybe on a wider scale.

By the end of the field visit the province manager of water distribution of Karakalpakstan was applying for a grant to fund his own support unit. The foundation he approached, Eurasia Foundation, had three areas of interest. To be more successful in the application for a grant, he raised in the proposal two issues which were considered by Eurasia to be worthy of support: training farmers as small entrepreneurs and social capital building at the local level. Even though, the province manager of water distribution in Karakalpakstan had not made the shift from top-down to participation, to increase his own security he had to acknowledge outside demand and possibilities. Not only do higher levels have large influence on changes but also outside possibilities influence the path of the changes.

11 Competing interests of stakeholders at the same level

Being located in different provinces the two province managers of water distribution were not competing with each other and even profited from sharing information about institutional arrangements and different experience. However, in relation to higher levels, such as the national or international level, they were competitors. They competed over the access to outside contacts and financial resources. Outside interest was positively related not only to their social position and influence, but also to the success of the water user associations in their region. Consequently they were encouraged to enhance the success story of their water user associations, or at least to promote their water user association experience as success story, so that it catches the attention of outside agencies.

The argument of competition on the provincial level, validates Jones Loung's reasoning. (see chapter three) However, it would be possible to reason that the districts were in competition at the local level. The better they performed as support units for water user associations or as water user associations, the more attention they would receive and the more likelihood to draw more projects and more funding. However, it seemed that at the current level at least the water user associations were not aware of these effects.

12 Supply and/or demand induced changes and the role of the provincial water elite

The evaluation indicates that the changes from state and collective farm to water user association were initially supply induced changes from the national level, which did not directly endanger the position of the province managers of water distribution. However, other changes of the system affected the status and security of the province managers of water distribution directly and more significantly. It seems that indirectly the changes from state and collective farm to water user association could have been perceived as a risk situation, or might have triggered other perceptions of risk. Indirectly, because the induced changes questioned the system of command management and initiated a system which gave farmers on the one hand responsibility for the irrigation and drainage system, and on the other hand freedom to decide about their representatives and how best to implement the role of responsibility. At the time of the study – 2000/2001 – the local level did not challenge the local existing structures.

The changes in the provincial allocation of water triggered different issues at the provincial level. The province manager of water distribution in Khorezm reinterpreted the changes as an opportunity for him to gain future security. He promoted the changes and manipulated the risk with the anticipated negative consequences into a positive situation. He initiated some training sessions and a support unit, furthermore it seems that he gained acceptance of the positive perception of the changes amongst local level stakeholders. As a result he gained opportunities to re-establish their lost power position and to gain more social recognition.

Arguably the supply induced changes did not satisfy different levels. These levels furthered the changes so that they could satisfy the demands at different levels and even create new demands, such as training at the local level. Supply induced changes initiated a demand for more change. Supply induced changes created demand induced changes.

13 Conclusion

The chapter focused on institutions at the provincial level responsible for water allocation and management. The institutional changes affecting the provincial level differed from the local level, and the local level changes did not have a direct influence on the provincial level. However, the stakeholders at the provincial level took an interest in the changes and started to influence the local levels more actively. At the beginning of the chapter it was shown that institutional changes affected the provincial level, and it was argued that this created a risk situation, or at least the actors involved interpreted it to be a risk situation or a situation of future uncertainty.

This risk situation was the trigger for the search by these supervisory stakeholders for other alternatives.

The perceived alternatives for the stakeholders at the provincial level, at the time of the research, involved changes at the local level. The provincial level officials made for themselves a window of opportunity, to leave their uncertain situation and to create new securities for themselves, at the expense of the local level. Because of their position within the institutions, the two province managers of water distribution could manipulate the local level events. For example, positive manipulation for the local farmers, but arguable in the case of Karakalpakstan and Khorezm; negative manipulation for the former state and collective farm managers; and in the case of Khorezm, negative manipulation for the district level, because they were excluded from the changes.

The influence of the province managers of water distribution shows that the argument of the theory chapter in terms of actors' influence on change is too simplistic. It is too simplistic because even though the theory acknowledges that institutions have different levels, and that these levels interact vertically, it did not take into account that these levels could interact diagonally or horizontally, or in even more complex ways. Hence, uncertainty can trigger complex actions for an actor. In comparison while the province manager of water distribution in Karakalpakstan went through the system, incorporating the district level, in Khorezm the province manager of water distribution changed the system by avoiding the district level and by approaching the local level directly. In addition, the example of the province manager of water distribution in Khorezm showed that his approach to create security was very dynamic, and that he was open and willing to co-operate with 'insiders' and 'outsiders' to achieve a level of security.

The study shows that from a holistic perspective, the borders between inside and outside become even more blurred. Arguably, the province managers of water distribution were insiders of the culture, of water management, of the hierarchy, however they were also outsiders to the local setting, to the local hierarchy, to farming, to the local culture on the former state and collective farm, in addition they had links to the outside, which even strengthened their outsider position, links to the national level, links to foreign experts, links to outside information. Taking a holistic and dynamic perspective there is no clear boundary between inside and outside. Hence, Douglas argument of real changes coming only from the outside, is too restricting and therefore highly questionable.

Chapter six: Conclusion

1 Findings

The data collected on the historical development of cotton production and the recent transformation from state and collective farms to farm organisations and water user associations in the two provinces indicates, that to get a better understanding of the local level shifts, one has to take a wider and more holistic and dynamic perspective. These perspectives have been missing in the studies on land reform and the establishment of water user associations. The main findings of the thesis can be summarised as:

1. Through land reforms parts of the elites or actors who belong to the networks of the elites established farms at the local level. These actors had and still have networks to higher levels than the local administrational levels and therefore have more power then the 'normal' local level actor. The new power actors at the local level change the existing balance of power at the local level. The local level as a whole becomes more influential and its bargaining power increases. This implies that it could take action and contradict the interests of the old local elite.
2. Land reform and the introduction of water user associations were in the case study area top-down changes. Through the land reforms parts of the elite moved to the bottom. Because of the new power relations at the local level, there has been a shift from top-down to bottom-up changes. The new local power actors have demands and have the bargaining power to ask for changes and to implement changes.

3. Whether the new power actors merge with the other local actors is context specific. In the case of the water user associations in Karakalpakstan and Khorezm province, merging took place if all the local actors shared the same resource (here the same irrigation canal) and it depended on the extent to which water scarcity was experienced. In the case where the water scarcity was very severe collective action did not occur, however, when the scarcity was less severe collective action took place.[28] Hence, the success of implementing changes is context specific.
4. The case study demonstrated that different higher administrative levels can function as facilitators for changes at lower levels. In Khorezm, the head of water distribution on the province level tried to facilitate changes on the local level. Facilitation took place in the form of knowledge management, which increased the transparency of the changes and helped to empower the local actors. The higher level facilitation rebalanced in the power relations at the local level between elites and other actors at the local level. The local elites and actors began to have a voice at higher administrative levels. In Karakalpakstan, the head of water distribution on the province level started to see himself as a facilitator.
5. Not every elite was identical. Elites had evolved through different structures. The evidence from the case studies demonstrated that the power of an elite can be based on the patronage system, the political party system or on the educational system. (other system have different elites, for example religious elites and market elites). Because the power of the elites was based on different structures, the different elites expressed varying levels of openness and adaptiveness to changes. In addition, not all changes were equally significant. Therefore, not every change challenged the existing structures

28 Collective action is not necessary if there is a enough water.

equally. Hence, it was possible to get support for different changes from different elites.

In addition to the main findings, there are a number of minor outcomes of the research, which are important for the general theory of institutional change, although less significant with respect to the case of land reform and institutional change in Karakalpakstan and Khorezm province. The minor research outcomes are:

6. Institutions change if the political and administrative context of the institutions change.
7. Institutions are less likely to function if exit opportunities exist.

Furthermore, the analysis of institutional change on the local administrational level in Karakalpakstan and Khorezm confirmed some of the general assumptions about institutions and institutional change.

8. The outcome of supply induced institutional change is not necessarily a social optimum.
9. Institutional innovations are path dependent.

In the following discussion these ideas will be further discussed in a holistic context.

2 Discussion

2.1 Politics and path dependency

As indicated in the third chapter, the history of agricultural production and water management in Uzbekistan was heavily influenced by politics throughout the nineteenth and twentieth century. Top down decisions were supposed to increase agricultural production. The high priority through the end of 19th and during the whole of the 20th century was on cotton production, with lower priority on rice production during the Soviet period and grain production since independence. During these times the policies of how to achieve an ever higher production has changed with ideology and politics. The target never did. Even though these policies either indirectly influenced or directly altered the institutions responsible for local water management, these institutions kept a certain pattern. This pattern was the system of patronage.

It seems that Uzbekistan and the other Central Asian republics are heavily influenced by the past policies of the Russians and the Soviet Union. The Soviet Union left behind a vulnerable and complex upstream-downstream water management arrangement, but also left downstream Uzbekistan and Turkmenistan with large agricultural tracts and water management structures which cannot be justified economically. The new republics have to decide whether they want to continue or to break with the past. However, a fast break would leave millions of people employed in agriculture without livelihoods. However, a continuation of old agricultural policies and even an extension of the current agricultural area would just prolong the insecure legacy and its problems.

At the local level, the findings from the former state and collective farms which were visited indicate that the local organisations and institutions are influenced by the past as well. It had been argued that even though the approaches during the Soviet Union aimed to undermine the system of patronage, the patronage system continued and was incorporated into the new system of state and collective farm and command management.

2.2 *Changing the path*

In the literature on the first land reform it is emphasised that former stakeholders from different levels are gaining access to land, while this is perceived to be negative, it is important to point out that the land distribution is only one part of the reform, a second part of the reform is the establishment of farm organisations and water user associations. While for the first reform the informal networks proved to be negative for the total population, in terms of land distribution and access to water these informal networks could prove to be positive when it comes to an increase of bargaining power of the local level. The reason for this is that these networks were not only directed to the former state and collective farm level but also to higher administrative levels such as the district or the province. This reasoning would confirm the first main finding, namely that because of the new power actors at the local level the existing balance of power at the local level changed. The local level as a whole became more influential and its bargaining power increased. The shift in bargaining power and in the articulation of new demands was the beginning of a wider structural shift. It was the beginning of a shift towards participation and from a top-down structure into a bottom-up structure. This shift also puts into question the argument of path dependency of institutions. The shift in power relations at the local level enabled local actors to take different paths and to be more critical about the existing path.

2.3 *Institutional change and the social optimum*

There is a direct link between supply or top-down induced institutional change and the social optimum. While it was noted in the second chapter that top-down induced changes are not the basis of a local social optimum, the case study put this assumption into a different perspective. Because, even though changes are not supplied at the social optimum, the consequences of the non-optimal supply can be positive. Hence, non-optimal supply induced institutional change does not have to be negative. A more holistic and dynamic approach is needed. The argument of social optimum covers only the supplied institutional innovation, but does not continue to look into the spin-offs and the triggered effects of the institutional innovation. The example from the water user association in Urgench shows that long-term spin-offs can be positive and have to be considered by evaluating the top-down institutional change.

In the case of the water user association in Urgench, the long-term effects of the non-optimal supplied institutional innovation were positive. One of the new farmers had links to the Province Department for Agriculture and Water Resources. Because of these links it was possible to create a balance against the power of the former state and collective farm manager. Hence, because of the network ties to higher administrative levels the local farmers could afford to be more critical and to demand more services from the farm organisation manager or from the water user association manager.

The involvement of the province managers of water distribution in Khorezm and in Karakalpakstan indicates that support units on different administrational levels were a positive development. These support units can be interpreted as facilitators of change. They not only managed knowledge in form of agricultural and irrigation expertise but were also an alternative power, which could re-balance the power of the local elite. Because knowledge is power, and because local farmers could get advice from these higher administrative levels and did not have to turn to the local elite, the influence of the local elite was diminished. In these new circumstances the support units at the local level lost the gate keeper function of managing information and knowledge.

2.4 The importance of context

As shown in chapter four the land reform brought new power actors on the local level. However, the arrival of new power actors does not imply that they join the other local actors in one group, to challenge the old local level elite. For example the Keigili water user association, experienced the arrival of new power actors who had access to land which utilised water from the main canal and not from a secondary canal. This implied that the water scarcity they experienced in the year 2000 and 2001 was less severe. Because of their favourable position it would have been negative for them to be critical of the farm organisation manager or the water user association manager. The position was similar in the Beruni water user association. Here, the top-down water distribution from the water user association manager was accepted, because new power actors could utilise their influence to receive more water. This implies that even though new power actors could increase the bargaining power of the whole community, but whether they totally merged their interests with the other local actors was context specific.

In the background chapter it was argued that during the Russian period the main emphasis was on creating economic incentives, while retaining the local institutional structures of water management. It was shown that the incentives changed the contextual setting in which the institutions operated. Because the contextual setting changed, the former formal constitutional rules could not anymore support the formal operational rules of water management. This situation led to an evolution of practices which further undermined the formal higher level rules. Hence, institutions changed if the political and administrative context of the institutions changed.

Having said this, because institutions colour the perceptions of reality, not every contextual change is registered, and if registered, the institutions also coloured the interpretations and solutions. It has been argued in this study that the water scarcity downstream was perceived as a natural drought, something which happens and is beyond the influence of the locals. However, it was not perceived to be permanent. Hence, applying for agricultural land made sense, because the interpretation of reality was that agriculture downstream was not more under threat than anywhere else in the country. Hence, the viability of the downstream institution ‘agriculture’ and irrigation were not questioned.

2.5 Boundaries and exit opportunities

The water user association in the Beruni district in Karakalpakstan showed that if a water user association is based on administrative boundaries and not on irrigation boundaries that collective action within the water user association is less likely. It is less likely because some members have exit opportunities, meaning that they have access to other water resources. Having access to exit opportunities implies that they do not have to comply with collective action. For the land reforms in Uzbekistan it indicates that the former state and collective farm boundary was not the right boundary for the new water user association.

As argued in the literature on common pool resources, exit opportunities weaken collective action institutions. The supply of new common pool resource institutions is less successful if exit opportunities exist. The new institution, which depends on collective choice decisions over one resource, would have to take into consideration that there is more than one resource, hence the actions by some members of the group will not directly affect the other members of the group, or only affect them at a different level. Hence, in this example there was not one common pool resource, but two.

3 Recommendations: from theory to practice

3.1 Support organisations and transparency

The evaluation of the interaction between the two different administrative levels shows that the results are more positive than when just looking at the local level at which the reforms take place. The reasoning indicated that the overall bargaining power on the local level increased, however it is still a question of how to challenge the bargaining power to the advantage of the whole community and not only to the advantage of the few. Hence, it seems that it would be necessary to establish organisations which can take over a similar role of the province manager for water distribution in Khorezm, so that the powerless have an alternative power to turn to, which is more powerful than the local power stakeholders. It would be necessary to establish support units which not only provide technical assistance but are also aware of social and cultural interactions and mechanisms.

However, the high inequality between the newly independent farmers only occurred because of unfair land distribution. If the land distribution would be more fair, there would be still a difference in bargaining power between the farmers, however, the difference would be smaller. To prevent unfair land distribution the bargaining power of the local level would have to be increased from the beginning. An increase in bargaining power could be achieved through more transparency during the reforms or the implementation of the reforms through neutral organisation, which are not tied to the old structures. A combination of these two approaches would imply that the reforms were not only implemented in a top-down approach but that there is also a bottom-up ingredient. Such approaches will make people more aware of the ongoing changes. Awareness could lead to more participation and therefore more control.

3.2 Quality not quantity

However, where donors are involved they want to see rapid changes and the establishment of large numbers of water user associations in a short time. Unfortunately the example of Kyrgyzstan shows such expectations to be unreasonable. The World Bank established criteria for success. Success is measured in seven 'milestones for irrigation system rehabilitation'. From the 250 established water user associations in Kyrgyzstan in 2001 only 40 had reached milestone four and only 20 milestone seven. This is an indication that the high expectations of the donor community have not lead to rapid sustainability. It shows that the process of institutional change, the change from the centrally managed farm to the participatory managed water user association is not just a change in name but a change of expectations and behaviour. It is a long process.

3.3 Not all constraints addressed

The study has focused on institutional change at the local level and the establishment of water user associations, however what has not been addressed in this study is whether the wider water management infrastructure and the command management system of agriculture is conducive to water user associations. These strategic issues been raised throughout the thesis, such as the focus on cotton over rice, with the impact of unequal water distribution or the economic and technical constraints resulting from the 'virgin land' policy of the Soviet Union. The evidence of arbitrary supply management of water and the continued command administrative approach to agricultural production suggests that the broader environment for sustainable water user associations is not in place. Overall the water user association are getting the water users more involved, in terms of taking over the costs of operation and maintenance, however these projects do not foster real local empowerment, because barriers to local empowerment are not addressed within these projects.

4 Summary

The case study of institutional change in water management at the local and provincial level in Uzbekistan has demonstrated that two different levels of the administrative structure interact and that these interactions can be very complex and dynamic. The findings challenge the current conceptual frameworks of institutions, institutional change, elites and their role during periods of change. While the current conceptual frameworks acknowledge different levels of institutions and actors, the interdependence and interactions between different institutional levels and actors at these different levels has not been deeply researched. In addition, the current frameworks focus on the identification of possible parameters for institutional change, however only an understanding of the interaction of these parameters in a dynamic environment can explain reality. Only the introduction of risk and uncertainty as primary motivation helped to interpret the actions of stakeholders in a dis-integrating institutional environment and gave insights into dynamic interactions between different institutional levels and actors.

Bibliography

Alchian, A. & Demsetz, H., 1973, 'The property rights paradigm', *The Journal of Economic History*, Vol. 33, pp. 16–27.

Alimov, S., 2000, 'Uzbekistan faces drought crisis', *Institute for War and Peace Reporting*, Issue No. 7: June 16, 2000.

Allan, J.A., 2001, *The Middle East water question: hydropolitics and the global economy*, London: I B Tauris.

Babu, S. & A. Tashmatov, 2000, 'Emerging issues and challenges for attaining food security in Central Asia', in S. Babu and A. Tashmatov (ed.), *Food policy reforms in Central Asia*, (Washington: International Food Policy Research Institute, DRAFT). (unpublished)

Baland, J-M & J-P Platteau, 1999, 'The ambiguous impact of inequality on local resource management', *World Development*, Vol. 27, No. 5, pp. 773–788.

Baumann, P., 1999, Kyrgyz Republic, agriculture area development project, social diagnosis. (unpublished)

Baumann, P., 1999, 'Kyrgyz Republic World Bank on-farm irrigation project social diagnosis: Osh and Jalalabad Regions,' SERA Report. (unpublished)

Berkes, F., (ed.) 1989, *Common property resources: ecology of community-based sustainable development*, London: Belhaven.

Berkes, F. & M.T. Farvar, 1989, 'Introduction and overview', in Berkes, F., (ed.), *Common property resources: ecology of community-based sustainable development*, London: Belhaven, Ch. 1, pp. 1–17.

Berkoff, D.J.W., 1994, 'Are collective farms water user associations? Land reform and irrigation management in Uzbekistan', paper submitted to the International Conference on Irrigation Management Transfer Wuhan, China: September 20–24, 1994 (DRAFT) (unpublished).

Binswanger, H.P. & V.W. Ruttan, 1978, *Induced innovation: technology, institutions, and development*, Baltimore: Johns Hopkins University Press.

Bloch, P., 1998, Synthesis of the January 1998 Central Asia Regional Planning Workshop, Basis Report, http://www.wiscinfo.wisc.edu/ltc/basprog2.html

Bloch, P. & A.A. Kutuzov, 2001, Rural factor market issues in the context of agrarian reform, Basis Report, http://www.wisc.edu/ltc/live/baseur0106a.pdf

Blowers, A. & P. Leroy, 1994, 'Power, politics and environmental inequality: a theoretical and empirical analysis of the process of 'peripheralisation,' *Environmental Politics*, Vol. 3, No. 2, pp. 197–228.

Bourdieu, P., 1990, *The logic of practice*, Polity Press.

Brooks, K. & Z. Lerman, 2000, Turkmenistan: an assessment of leasehold-based farm restructuring. (unpublished)

Burger, R., 1998, 'Water User Associations in Kazakhstan: an institutional analysis,' *Environment Discussion Paper* No. 45. (web-publication)

Carlisle, D.S., 1991, 'Power and politics in Soviet Uzbekistan: from Stalin to Gorbachev', in Fierman, W., 1991, *Soviet Central Asia: the failed transformation*, Oxford: Westview Press.

Carter, J., 1982, *Keeping faith: memories of a president*, New York: Bantam Books Cashdan, E.A. (ed.), 1990, *Risk and uncertainty in tribal and peasant economies*, Boulder: Westview Press.

Clem, R.S., 1997, 'The new Central Asia: prospects for development', in Bradshaw, M.J., 1997, *Geography and transition in the post-soviet republics*, New York

Coleman, J.S. 1990, 'The emergence of norms', in Hechter, M., K.D. Opp, & R. Wippler, 1990, *Social institutions, their emergence, maintenance and effects*, New York: Walter de Gruyter.

Critchlow, J., 1991a, *Nationalism in Uzbekistan*, Boulder: Westview Press.

Critchlow, J., 1991b, 'Prelude to 'independence': how the Uzbek party apparatus broke Moscow's grip on elite recruitment', in

Fierman, W., 1991, *Soviet Central Asia: the failed transformation*, Oxford: Westview Press.
DAI, 1996, An environmental profile of the Republic of Uzbekistan, Report, Prepared for the ADB
Das Gupta, M., 2001, State-community synergies in development: laying the basis for collective action, World Bank, Development Research Group
De Janvry, A., E. Sadoulet & E. Thorbecke, 1993, 'Introduction', *World Development*, Vol. 21, No. 4, pp. 565–575
DFID, 2001, Addressing the water crisis – healthier and more productive lives for poor people.
Douglas, M., 1986, *How institutions think*, London: Routledge & Kegan Paul.
Douglas, M. and A. Wildavsky, 1982, *Risk and culture: an essay on the selection of technical and environmental*, Berkeley: University of California Press.
Downs, A., 1957, *An economic theory of democracy*, New York: Harper & Row.
Eckert, J. & G. Elwert, 2000, Land tenure in Uzbekistan, FU-Berlin, Socialanthropologische Arbeitspapiere Nr.86.
Edwards, M., 2001, 'Enthusiasts, tacticians and sceptics: the World Bank, civil society and social capital.' (web publication)
Egamberdi, N. & P. Gordon, A. Ilkhamov, D. Kandiyoti, and J. Schoeberlein: Uzbekistan agriculture enterprise restructuring and development program, World Bank: Best Practice Social Assessments:
http://WBLN0018.Worldbank.org/Networks/ESSD/icdb.nsf/d4856f112e805df4852566c9007c27a6/3e728b8b77815a7185256832005484c3/$FILE/chapter6a.pdf
Ensminger, J. & J. Knight, 1997, 'Changing social norms: common property, Bridewealth, and Clan Exogamy', *Current Anthropologist*, Vol. 38, Issue 1, pp. 1–24.
Feeny, D., 1988, 'The demand for and supply of institutional arrangements' in Ostrom, V., D. Feenly & H. Picht, 1988, *Rethinking institutional analysis and development*, San

Francisco: International Centre for Economic Growth, Chapter 6, pp. 159–209.
Garston, Neil (ed.), 1993, *Bureaucracy: three paradigms*, London: Kluwer.
Gibbs, C.J.N. & D.W. Bromley, 1989, 'Institutional arrangements for management of rural resources: common-property regimes', in Berkes, F., (ed.), 1989, *Common property resources: ecology of community-based sustainable development*, London: Belhaven.
Glantz, M. (ed.), 2002, Workshop Report of informal planning meeting: water, climate, and development issues in the Amu Darya basin, 18–19 June 2002, The Franklin Institute, Philadelphia, Pennsylvania. (web-publication)
Glazovsky, N.F., 1995, 'The Aral Sea basin', in Kasperson, J.X., R.E. Kasperson, and B.L. Turner II, (ed.), 1995, *Regions at risk: comparisons of threatened environments*, Tokyo; New York :United Nations University Press.
Haas, P.M., 1994, 'Do regimes matter? Epistemic communities and mediterranean pollution control', in Kratochwil, F. & E.D. Mansfield, 1994, *International organization: a reader*, New York: HarperCollins College Publishers, pp. 129–139.
Hajer, M, A., 1995, *The politics of environmental discourse: ecological modernization and the policy process*, Oxford: Clarendon Press.
Hardin, G., 1968, 'The tragedy of the commons', *Science*, Vol. 162.
Hechter, M., K.D. Opp, & R. Wippler, 1990, *Social institutions: their emergence, maintenance and effects*, Berlin; New York: Walter de Gruyter.
Hechter, M., 1990, 'The emergence of cooperative social institutions,' in Hechter, M., Opp, K.D. & R. Wippler, 1990, *Social institutions: their emergence, maintenance and effects*, Berlin; New York: Walter de Gruyter.
Hobley, M. & K. Shah, 1996, 'What makes a local organisation robust? Evidence from India and Nepal', *ODI Natural Resource Perspective* No. 11.

Hobley, M. & D. Shields, 2000, 'The reality of trying to transform structures and process: forestry in rural livelihoods', ODI, *Working Paper* 132.

Hobsbawm, E. & T. Ranger, (ed.), 1983, *The invention of tradition*, Cambridge: Cambridge University Press.

Homer-Dixon, T., 1995, 'The ingenuity gap: can poor countries adapt to resource scarcity?', *Population and Development Review*, 21, no. 3 pp. 1–26 (web publication: http://www.library.utoronto.ca/pcs/).

Hutchens, A.O., S.H. Johnson III & J.R. Strickland, 2000, 'Dashowuy irrigation management study', EPIC-report, Central Asia Mission U.S. Agency for International Development.

Ilkhamov, A., 1998, 'Shirkats, dekhqon farmers and others: farm restructuring in Uzbekistan', *Central Asian Survey* Vol. 17(4) pp. 539–560.

Johnson III, S.H., 1999, 'WUA specialist report, on-farm irrigation project government of Kyrgyz Republic.' (unpublished)

Jones Luong, P., 2000, 'Sources of institutional continuity: the soviet legacy in Central Asia' http://www.duke.edu/web/licep/2/luong/luong.pdf

Keohane, R.O., 1994, 'International institutions: two approaches' in Kratochwil, F. and E.D. Mansfield, *International organization: a reader*, New York: HarperCollins College Publishers.

Keohane, R.O., 1983, 'The demand for international regimes', in Krasner, S.D., *International regimes*, Ithaca: Cornell University Press.

Kotlyakov, V.M., 1992, 'Concept for preserving and restoring the Aral Sea and normalizing the ecological, public health and socioeconomic situation in the Aral region', *Post Soviet Geography*, Vol. 33, No. 5, pp. 283–295.

Kotlyakov, V.M., 1991, 'The Aral Sea basin: a critical environmental zone,' *Environment*, Vol.33 no.1, p. 4–9 & 36–38.

Kropp, B., M. Lundell, J. Lampietti, B. Shamsiev, 'Rural welfare chapter – concept summary and outline living standards assessment, Uzbekistan'. (draft version 2002) (unpublished)

Khusanov, R., 2000, 'Policy reform and achieving food security in Uzbekistan', in Babu, S. & A. Tashmatov, (ed.), 2000, *Food policy reforms in Central Asia*, Washington: International Food Policy Research Institute (DRAFT) (unpublished).

Lerman, Z., J. Garcia-Garcia, & D. Wichels, 1996, 'Land and water policies in Uzbekistan', *Post-Soviet Geography and Economics*, Vol.37, No.3, pp. 145–174.

Lerman, Z., & K. Brooks, 1998, 'Land reform in Turkmenistan', in Wegren, S.K.(ed.): *Land reform in the former Soviet Union and Eastern Europe*, London: Routledge.

Lipovsky, I., 1995, 'The Central Asian cotton epic', *Central Asian Survey* Vol.14.

Marcus, G.E. (ed.), 1983, *Elites: ethnographic issues*, Albuquerque: University of New Mexico Press.

Matley, I.M., 1967, 'Agricultural development' in Allworth, E. (ed.), *Central Asia: A century of Russian rule*, New York, London: Columbia University Press.

McKee, M. and M. Curtin, 1996, 'An environmental profile of the Republic Uzbekistan,' DAI.

McKinney, D.C., 1997, 'Sustainable water management in the Aral Sea Basin,' http://www.ce.utexas.edu/prof/McKinney/papers/aral/AralSus.html

Meyer, J.W., J. Boli, & G.M. Thomas, 1994, 'Ontology and rationalisation in the Western cultural account', in Scott, R.W. & J.W. Meyer, 1994, *Institutional environments and organizations*, London; Thousand Oaks, California: SAGE Publications.

Micklin, P.P., 1992, 'The Aral Crisis: introduction to the special issue,' *Post Soviet Geography*, Vol. 33, No. 5, pp. 269–282.

Mosse, D., 1997, 'The symbolic making of a common property resource: history, ecology and locality in a tank-irrigated landscape in South India,' *Development and Change*, Vol. 28, pp. 467–504.

Mueller, D.C., 1989, *Public choice II*, Cambridge: Cambridge University Press.

Muller, M. 2002, Regional water management in Central Asia. (unpublished)

Murphy, S., 1999, Trade and food security: an assessment of the Uruguay round agreement on agriculture, *CIIR discussion*, London.

Nabli, M.K. & J.B. Nugent, 1989, 'The New Institutional Economics and its applicability to development', *World Development*, Vol.17, pp. 1333–134.

Niskanen, W.A., 1994, *Bureaucracy and public economics*, Aldershot, Hants, England; Brookfield, Vt., USA : E. Elgar Pub.

North, D.C., 1990, *Institutions, institutional change and economic performance*, Cambridge: Cambridge University Press.

OCHA, 2001, 'Uzbekistan – drought', OCHA Situation report No. 2.

O'Hara, S. L., 2000, 'Lessons from the past: water management in Central Asia,' *Water Policy*, No. 2, pp. 365–384.

Ohlsson, L., 1999, *Environment, scarcity, and conflict: a study of Malthusian concerns*, Göteborg: Department of Peace and Development Research, Göteborg University.

Olson, M., 1971, *The logic of collective action: public goods and the theory of groups*, Cambridge, Mass.; London: Harvard University Press.

O'Neill, G., forthcoming, 'Land and water reform in the 1920s: agrarian revolution or social engineering?' (unpublished)

Opp, K.D., 1999, 'Contending conceptions of the theory of rational action,' *Journal of theoretical politics*, Vol. 11, pp. 171–202.

Ortner, Sherry B., 1989, *High religion: a cultural and political history of Sherpa Buddhism*, Princeton: Princeton University Press.

ORYZA Market Report, 2000, Uzbekistan, Nov 29.

Ostrom, E., 1990, *Governing the commons: the evolution of institutions for collective action*, Cambridge: Cambridge University Press.

Ostrom, E., R. Gardener and J. Walker, 1994, *Rules, games, and common-pool resources*, Ann Arbor: University of Michigan Press.

Ostrom, V., D. Feenly & H. Picht, 1988, *Rethinking institutional analysis and development*, San Francisco: International Centre for Economic Growth.

PA Consortium Group & PA Consulting, 2002, 'Transboundary Water and Related Energy Cooperation for the Aral Sea Basin Region of Central Asia'. (Funded by: U.S. Agency for Internati9onal Development Regional Mission for Central Asia Office of Energy and Water)

Pallot, J., 1995, 'Agriculture and rural development', in Shaw, D.J.B. (ed.), 1995, *The post-soviet republics: a systematic geography*, Halow: Longman Scientific & Technical.

Papadakis, E., 1996, *Environmental politics and institutional change*, Cambridge: Cambridge University Press.

Pierce, R.A., 1960, *Russian Central Asia 1867–1917*, Berkeley and Los Angeles: University of California Press.

Pomfret, R., 2000, 'Agrarian reform in Uzbekistan: why has the Chinese model failed to deliver?' *Economic development and cultural change*, University of Chicago.

Putnam, R.D., 1993, 'The prosperous community', *The American Prospect*, Vol. 4, Issue 13.

Putnam, R. D., 1976, *The comparative study of political elites*, London

Reisner, M, 1984, *Cadillac desert*, New York: Penguin Books.

Renger, J., 1998, 'The institutional framework of water management in the Aral Sea basin and Uzbekistan', European Union – TACIS Programme. (unpublished)

Robbins P., 2000, 'The rotten institution: corruption in natural resource management,' Political Geography, Vol. 19, pp. 423–443.

Romanova, O., 2001, 'Uzbek drought', Institute for War and Peace Reporting, Issue No. 66.

Rumer, B.Z., 1989, *Soviet Central Asia: 'a tragic experiment'*, London: Unwin Hyman.

Rumer, B.Z., 1991, 'Central Asia's cotton economy and its costs', in W. Fierman (ed.), 1991, *Soviet Central Asia, the failed transformation*, Boulder, Westview Press.

Sahlins, M., 1987, *Island of History*, Chicago: University of Chicago Press.

Scott, R.W. & J.W. Meyer, 1994, *Institutional environments and organisations*, London; Thousand Oaks, Calif: SAGE Publications.

Shanmugaratnam, N., T. Vedeld, A. Mossige, & M. Bovin, 1992, 'Resource management and pastoral institution building in the West African Sahel' *World Bank Discussion Papers*, No. 175.

Shepsle, K.A., 1989, 'Studying institutions', *Journal of theoretical politics*, Vol. 1, pp. 131–147.

SIC ICWC, 1999, 'Institutional, technical and financial issues facing the irrigation sector in the Central Asian Republics', Consultant Report.

Sokolov, V., 1999, 'Integrated water resource management in the Republic of Uzbekistan,' *Water International*, Vol. 24, No. 2.

Sperber, D. and D. Wilson, 1986, *Relevance: communication and cognition*, Oxford: Blackwell.

Spoor, M., 1995, 'Agrarian transition in former soviet Central Asia: A comparative study of Kyrgyzstan and Uzbekistan', *The Journal of Peasant Studies*, Vol. 23, No. 1, pp. 46–63.

Spoor, M., 1999, 'Agrarian transition in former soviet Central Asia: A comparative study of Kazakhstan, Kyrgyzstan and Uzbekistan', *ISS Working Paper* No. 298.

Swyngedouw, E., 1999, 'Sustainability, risk and nature: the political ecology of water in advanced countries', Proceedings of a workshop convened in the University of Oxford, 15–17 April 1999. (unpublished)

TACIS, 1997, 'Water resource management and agricultural production in the Central Asian Republics,' WARMAP project.

Tashmatov, A., 2000, 'Food policy and natural resource issues within the context of sustainable agricultural development: The Republic of Uzbekistan', in Babu, S. and A. Tashmatov, (ed.)

2000, *Food policy reforms in Central Asia*, Washington: International Food Policy Research Institute, DRAFT. (unpublished)

The Times of Central Asia, 2001, UN FAO Experts comment on Central Asian regional drought, July 1.

Thurman, M., 1999, 'The 'command-administrative system' in cotton farming in Uzbekistan 1920s to present,' *Papers on Inner Asia*, No. 32.

UNESCO report, 2000, 'Water related vision for the Aral Sea basin.'

UzbekWorld.com, 2001, 'Uzbek cotton target down, due to the drought,' September 12.

Wade, R., 1988, *Village republics: economic conditions for collective action in south India*, Cambridge: Cambridge University Press.

Wade, R., 1987, 'The management of common property resources: finding a cooperative solution.' *The World Bank Research Observer*, 2(2), pp. 219–234.

WARMAP Project 1995a, Phase 1 – project preparation reports, Volume 6, Legal and institutional aspects

WARMAP Project 1995b, Phase 1 – project preparation reports, Executive summary

Wegerich, K., 2001, 'Water User Associations in Uzbekistan and Kyrgyzstan', SOAS Water Issue Group, *Occasional Paper* No 32.

Wegerich, K., 2002a, 'Coping with dis-integration of a river-basin management system: multi-dimensional issues in Central Asia', (presented at SOAS May 2002). (*Water Policy*, accepted) (unpublished)

Wegerich, K., 2002b, 'Institutional challenges to WUA establishment in Uzbekistan,' presented at ICID March 2002. (unpublished)

Wegerich, K., 2002c, 'Natural drought or human made water scarcity in Uzbekistan?' *Central Asia and the Caucasus*, No. 2 (14).

Wegerich, K., 2003, 'Network structures and water distribution in two districts in Khorezm, Uzbekistan', presented at the Conference: Governing Environmental Flows, Wageningen University &

ISA (RC24), Wageningen, Netherlands, 13 & 14.06.2003 (unpublished)
Wegerich, K. & K. Bektemirov, 2001, 'Food self-sufficiency of food sovereignty for Uzbekistan: an analysis in context of water scarcity', *OCEES Research Paper* No. 23.
Weinthal, E., 1998, *Making or braking the state*, Ph.D. Thesis Columbia University. (unpublished)
Weinthal, E., 2002, *State making and environmental co-operation*, MIT Press.
Welsh, W. A., 1979, *Leaders and elites*, London.
Winiecki, J., 1996, 'Why economic reforms fail in the soviet system', in Alston, L.J., T. Eggertsson & D.C. North, 1996, *Empirical studies in institutional change*, Cambridge: Cambridge University Press, Ch. 2 pp. 63–91.
Woolcock, M., 1999, 'Managing risk, shocks, and opportunity in developing economies: the role of social capital', The World Bank, Development Research Group.
World Bank, 1998, 'Turkmenistan, Dashkhouvuz: regional development priorities' Main report, Volumes I and II, Report No. 17737-TM, Washington, DC: World Bank.
World Bank, 1996, 'Fundamental provision of water management strategy in the Aral Sea basin.'
Young, O.R., 1989, *International co-operation: building regimes for natural resources and the environment*, Ithaca; London: Cornell University Press.
Zonn, I., 1999, 'The impact of political ideology on creeping environmental changes in the Aral Sea basin', in M. H. Glantz, *Creeping environmental problems and sustainable development in the Aral Sea basin*, Cambridge University Press.

Appendix

List of Interviewees

Mr. Alisher Ilkhanov, Director, Centre of Social Research, 03. July 2000

Dr. R. Ikramov, General Director, SANIIRI, 03. July 2000

Dr. T. Bekmuratov, Expert, SANIIRI, 03. –06. July 2000

Mr. Rashid Koshekov, Head of Water Distribution, Autonomous Republic Karakalpakstan, 24. July–03. August 2001

Mr. Babajanov, Head of Water Distribution, Khorzm Province, 06. July 2000

Dr. Vadim Sokolov, Deputy Director, SIC ICWC, 03. July 2000

Dr. Pinkhasov, Expert, SIC ICWC, 03. July 2000

Dr. Nagmet Aimbetov, Director, Academy of Sciences of the Republic Uzbekistan, Karakalpak branch, The Institute of the Aral sea region's socio-economic problems, 24. July–03. August 2001

Dr. Kuatbay Bektemirov, Department Head, Academy of Sciences of the Republic Uzbekistan, Karakalpak branch, The Institute of the Aral sea region's socio-economic problems, 24. July–10. August 2001

Mr. Wob Groot, Water User Association Expert, 13. August–4. September 2001

Mr. Jeremy Berkoff, Water management expert, 27. June 2003

Mr. Suleimenov, Head of Water Distribution, Kungrad district, Autonomous Republic Karakalpakstan, 26. July 2001

Interviewees in different Water User Associations

(Interviewees have not been asked their names)

Keigili District, Autonomous Republic Karakalpakstan,
27. July 2001
District manager for water distribution
Former state and collective farm manager
Head of the Water User Association
Farmer 1, in meeting
Farmer 2, in meeting
Farmer 3, in meeting
Farmer 4, on the field
Farmer 5, on the field

Beruni District, Autonomous Republic Karakalpakstan,
28. July 2001
District manager for water distribution
Former state and collective farm manager
Head of the Water User Association
Farmer 1, in meeting
Farmer 2, in meeting
Farmer 3, in meeting
Farmer 4, in meeting
Farmer 5, on the field
Farmer 6, on the field

Khiva District, Khorezm Province,
06. July 2000
Former state and collective farm manager
Head of the Water User Association
Farmer 1, in meeting
Farmer 2, in meeting

Farmer 3, in meeting
Farmer 4, in meeting
Farmer 5, in meeting
Farmer 6, on the field
Farmer 7, on the field
Farmer 8, on the field

Urgench District, Khorezm Province,
28. July 2001
Former state and collective farm manager
Head of the Water User Association
Farmer 1, in meeting
Farmer 2, in meeting
Farmer 3, in meeting
Farmer 4, in meeting
Farmer 5, on the field

Syrdarya Province,
04. July 2000
Head of the Water User Association
Farmer 1, in meeting
Farmer 2, in meeting
Farmer 3, in meeting
Farmer 4, in meeting
Farmer 5, in meeting
Farmer 6, on the field
Farmer 7, on the field
Farmer 8, on the field